T0258792

$$pV = nRT$$

$$\frac{pV}{nRT} = 1 + B'p + C'p^2 \dots$$

$$-\left(\frac{\partial S}{\partial p}\right)_T = \left(\frac{\partial V}{\partial T}\right)_p$$

The Newman Lectures on Thermodynamics

The Newman Lectures on Thermodynamics

John Newman
Vincent Battaglia

JENNY STANFORD
PUBLISHING

Published by

Jenny Stanford Publishing Pte. Ltd.
Level 34, Centennial Tower
3 Temasek Avenue
Singapore 039190

Email: editorial@jennystanford.com
Web: www.jennystanford.com

British Library Cataloguing-in-Publication Data
A catalogue record for this book is available from the British Library.

ISBN 978-981-4774-26-0 (Hardcover)
ISBN 978-1-315-10861-2 (eBook)

Contents

Introduction

This material serves both as a review and as an introduction. Consequently, the readers should expect to learn this material later, and it may be useful to return to the Introduction to give them a better overall view of the subject of thermodynamics. Hopefully this review will inspire them to get on with learning the thermodynamics in the following chapters.

For a gas mixture obeying the truncated virial equation,

$$\frac{p\tilde{V}}{RT} = 1 + B'p, \tag{1}$$

where

$$B' = \sum_i \sum_j y_i y_j B'_{i,j}(T) \text{ and } B'_{i,j} = \frac{B_{i,j}}{RT}, \tag{2}$$

the Gibbs function becomes

$$\tilde{G} = RT \ln p + RTB'p + \sum_i y_i \mu_i^*(T) + RT \sum_i y_i \ln y_i. \tag{3}$$

This formulation is important because all thermodynamic properties can be derived when the Gibbs function is expressed as a function of T, p, and n_i. This equation applies to real gases up to 10 or 20 atm and gives explicitly the dependence on T, p, and n_i. You can set $B'_{i,j}$ equal to zero if you want to recover equations for an ideal gas, but the equations retaining $B'_{i,j}$ are valuable because they reveal how real gases may or may not approach ideal-gas behavior as p approaches zero. (For example, $(\partial H / \partial p)_T$ does not approach zero for a real gas but is identically zero for an ideal gas.) On the other hand, you may want to substitute more sophisticated equations of state to give accurate treatment at pressures greater than 10 or 20 atm.

Review Questions

1. Suppose that you have been given the Gibbs function G in terms of temperature, pressure, and mole numbers as independent

variables. Describe briefly, with equations, how you would obtain the following quantities:

   a. Heat capacity, C_p.
   b. Compressibility factor.
   c. Equilibrium composition if it is perceived that a single chemical reaction is possible among the constituents present.

2. Reflect on the possibility of working in reverse order to infer the Gibbs function from data on the heat capacity and compressibility factor for a nonreacting system. Do you also need the results of reversible mixing experiments (such as what the Gibbs mixing rule tells you)?

The chemical potential of a component in a gas mixture can be expressed as

$$\mu_i = \mu_i^*(T) + RT \ln(py_i\phi_i), \tag{4}$$

an equation that merely *defines* the fugacity coefficient ϕ_i. Manipulation of G from Eq. 3 according to the definition of the chemical potential

$$\mu_i = \left(\frac{\partial n\tilde{G}}{\partial n_i}\right)_{\substack{T,p,n_j \\ j \neq i}} \tag{5}$$

and comparison of the result with Eq. 4 permit us to write

$$\ln \phi_i = -B'p + 2p \sum_j y_j B'_{i,j}. \tag{6}$$

For a binary mixture, this takes the explicit form

$$\ln \phi_1 = p\left[y_1(2-y_1)B'_{1,1} + 2y_2^2 B'_{1,2} - y_2^2 B'_{2,2}\right]. \tag{7}$$

Both these expressions show that ϕ_i approaches 1 as p approaches zero. Also, one sees that $\phi_i = 1$ for an ideal-gas mixture (where $B'_{i,j} = 0$).

From Eq. 4, the entropy is

$$\tilde{S} = -\left(\frac{\partial \tilde{G}}{\partial T}\right)_{p,n_i} = -R \ln p - Rp\frac{\partial(B'T)}{\partial T} - \sum_i y_i \frac{d\mu_i^*}{dT} - R\sum_i y_i \ln y_i. \tag{8}$$

The heat capacity then follows as

$$\tilde{C}_p = T\left(\frac{\partial \tilde{S}}{\partial T}\right)_{p,n_i} = -RTp\frac{\partial^2 (B'T)}{\partial T^2} - T\sum_i y_i \frac{d^2 \mu_i^*}{dT^2}. \qquad (9)$$

If we let p approach zero, this equation then leads to the result

$$\frac{d^2 \mu_i^*}{dT^2} = -\frac{\tilde{C}_{pi}^*}{T}, \qquad (10)$$

where $\tilde{C}_{pi}^*$ is the low pressure limit of the molar heat capacity of a pure component. Integration of Eq. 10 to give μ_i^* from data on $\tilde{C}_{pi}^*$ requires two integration constants to be specified. These correspond to the primary reference states for both entropy and enthalpy for each substance. For reacting systems, these two integration constants can be related back to primary reference states for the elements.

Chemical Equilibrium

Minimization of the Gibbs function for a reacting system under constraints of constant temperature and pressure leads to the general equilibrium condition:

$$\sum_i v_{i,\ell} \mu_i = 0, \qquad (11)$$

where $v_{i,\ell}$ denotes the stoichiometric coefficient of species i in reaction ℓ. A separate equation of this form applies for each *independent* chemical reaction.

Substitution of Eq. 4 *immediately* yields

$$K_\ell = \prod_i (y_i p \phi_i)^{v_{i,\ell}}, \qquad (12)$$

where

$$K_\ell = \exp\left(-\frac{\Delta G^*}{RT}\right) \qquad (13)$$

and

$$\Delta G^* = \sum_i v_{i,\ell} \mu_i^*. \qquad (14)$$

K_ℓ is called the thermodynamic equilibrium constant for the reaction and is a function of temperature only. ΔG^* is called the standard

Gibbs function change for the reaction. It should perhaps be noted that ΔG^* and μ_i^* are collections of integration constants or secondary-reference-state quantities and not actually the Gibbs function or the chemical potentials themselves. In this regard, the notation is confusing.

Based on the fundamental equation

$$\frac{d \ln K}{dT} = \frac{\Delta H^*}{RT^2},$$

we can work out the temperature dependence of the equilibrium constant:

$$\ln \frac{K}{K(T_0)} = -\frac{\Delta H^*(T_0)}{R}\left(\frac{1}{T} - \frac{1}{T_0}\right) + \frac{\Delta a}{R}\left(\ln \frac{T}{T_0} + \frac{T_0}{T} - 1\right)$$

$$+ \frac{(T-T_0)^2}{2RT}\left(\Delta b + \frac{\Delta c}{TT_0^2} + \frac{T+2T_0}{3}\Delta\gamma\right), \tag{15}$$

where the temperature dependence of the heat capacity at low pressure is expressed as

$$\tilde{C}_{pi}^* = a_i + b_i T + c_i / T^2 + \gamma_i T^2, \tag{16}$$

and

$$\Delta a = \sum_i v_{i,\ell} a_i, \quad \Delta H^* = \sum_i v_{i,\ell} \tilde{H}_i^*, \text{ etc.} \tag{17}$$

Equations 12 and 15 provide a rather complete account of the temperature and pressure dependence of the equilibrium composition in a system of chemically reacting gases. The description is rather good even if one does not take the trouble to estimate the fugacity coefficients and takes instead $\phi_i = 1$. The results are of industrial importance because they allow the prediction of the maximum possible conversion in a chemical reactor. Furthermore, they define the proper driving force tending to propel reactions toward equilibrium when the kinetics of the reactions are used in the design of the reactor. Fundamental data for treating a variety of important systems at various temperatures and pressures can be summarized on a single sheet, as in Table 1.

In addition to the equilibrium relations (Eq. 12) and the means for calculating the equilibrium constant (Eq. 15), it is necessary to deal with material balances or stoichiometry of the reactions in order to complete the determination of the equilibrium composition.

Table 1 Thermodynamic data for treating chemical equilibria in gas-phase systems.

	$\widetilde{H}_i^*$ $\dfrac{\text{kcal}}{\text{mol}}$	μ_i^* $\dfrac{\text{kcal}}{\text{mol}}$	a $\dfrac{\text{cal}}{\text{K-mol}}$	$10^3 b$ $\dfrac{\text{cal}}{\text{K}^2\text{-mol}}$	$10^{-5} c$ $\dfrac{\text{cal-K}}{\text{mol}}$	$10^6 \gamma$ $\dfrac{\text{cal}}{\text{K}^3\text{-mol}}$
H_2	0	0	6.52	0.78	0.12	—
CO	−26.416	−32.78	6.79	0.98	−0.11	—
CH_3OH	−47.96	−38.72	4.394	24.274	—	−6.855
CO_2	−94.051	−94.254	10.57	2.10	−2.06	—
H_2O	−57.796	−54.634	7.30	2.46	—	—
CH_4	−17.88	−12.13	5.65	11.44	−0.46	—
N_2	0	0	6.83	0.90	−0.12	—
NH_3	−11.02	−3.94	7.11	6.00	−0.37	—
NO	21.57	20.69	7.09	0.92	−0.14	—
NO_2	7.93	12.26	10.07	2.28	−1.67	—
N_2O	19.61	24.90	10.92	2.06	−2.04	—
S_2	30.68	18.96	8.72	0.16	−0.90	—
H_2S	−4.93	−8.02	7.81	2.96	−0.46	—
SO_2	−70.944	−71.748	11.04	1.88	−1.84	—
SO_3	−94.58	−88.69	13.90	6.10	−3.22	—
HCHO	−28	−27	—	—	—	—
HNO_3	−32.28	−17.87	12.75	—	—	—

Note: $\widetilde{H}_i^*$ and μ_i^* are given at 298.15 K.

Part A
PURE SUBSTANCES

Chapter 1

Thermodynamics

Thermodynamics deals with heat and the interconversion of different forms of energy. An important application, which motivated much of the early development, is the production of electric power from thermal energy, the thermal energy having its origin in the chemical energy of fossil fuels or the nuclear energy of uranium, or perhaps direct solar energy. Thermodynamics is of widespread interest, courses being found in departments of mechanical engineering, physics, chemistry, and chemical engineering as well as, perhaps, biochemistry, electrical engineering, and materials science.

The special interest of chemical engineers is in chemical equilibria and phase equilibria. The former are pertinent because we are involved in the synthesis of chemicals via chemical reactions; equilibrium data are necessary in order to predict under what conditions the reaction will proceed and what yield can be expected. The latter are of interest because chemical engineers spend at least as much time in the physical separation of their product chemicals as they do producing them by chemical reactions. Thus, distillation, liquid extraction, and gas absorption, as well as humidification and evaporative cooling, require a knowledge of phase equilibria for their rational design.

From the nature of these processes, both chemical and physical, you can perceive that chemical engineers are most involved with the

The Newman Lectures on Thermodynamics
John Newman and Vincent Battaglia
Copyright © 2019 Jenny Stanford Publishing Pte. Ltd.
ISBN 978-981-4774-26-0 (Hardcover), 978-1-315-10861-2 (eBook)
www.jennystanford.com

thermodynamics of mixtures or, more properly, of multicomponent solutions. This training will continue in Part B, after the development in Part A of a background in single-component thermodynamics with emphasis on topics of interest to chemical engineers.

Thermodynamics does not deal with the rates at which these chemical and physical processes will occur, although it does define the driving forces for the processes. Thus, the final design of a chemical reactor requires a study of the kinetics of chemical reactions, and the design of a heat exchanger or an absorption column requires a knowledge of interphase transfer. One thing that you can do with thermodynamics alone is to "design" or determine the number of *equilibrium* stages required for a separation process—an example is the McCabe–Thiele diagram for calculating the number of equilibrium stages in a distillation column.

On a more prosaic level, thermodynamics is a procedure for recording experimental data in the simplest possible form. Instead of saying that so much heat was added to raise the temperature of 1.73 kg of water from 23.2° to 24.7°, we report instead the heat capacity, that is, we divide by the mass and by the temperature rise. This makes it easier to apply the result in other situations.

Along this same line of thought, thermodynamics also tells us how to avoid duplication in the measurement and recording of data. Suppose that we are asked to predict the final pressure when 1 kg of mercury is heated from 25°C to 30°C at constant volume, starting at 1 atm. We are given the isothermal compressibility

$$k = -\frac{1}{V}\left(\frac{\partial V}{\partial p}\right)_T = 3.9 \times 10^{-6}\, \text{atm}^{-1} \qquad (1.1)$$

and the coefficient of thermal expansion

$$\beta = \frac{1}{V}\left(\frac{\partial V}{\partial T}\right)_p = 1.8 \times 10^{-4}\, K^{-1}. \qquad (1.2)$$

What we need for the calculation is $(\partial p/\partial T)_V$, but this is not given directly. If pressure is a function of temperature and volume, we can write the total differential as

$$dV = \left(\frac{\partial V}{\partial p}\right)_T dp + \left(\frac{\partial V}{\partial T}\right)_p dT. \qquad (1.3)$$

At constant volume, or $dV = 0$, we then have

$$0 = \left(\frac{\partial V}{\partial p}\right)_T \left(\frac{\partial p}{\partial T}\right)_V + \left(\frac{\partial V}{\partial T}\right)_p \qquad (1.4)$$

or

$$\left(\frac{\partial p}{\partial T}\right)_V = -\frac{(\partial V / \partial T)_p}{(\partial V / \partial p)_T} = \frac{\beta V}{kV} = \frac{\beta}{k}. \qquad (1.5)$$

Thus,

$$p_2 = p_1 + \int_{T_1}^{T_2} \left.\frac{\beta}{k}\right|_{\text{const } V} dT = 1 + \frac{1.8 \times 10^{-4}\,\text{K}^{-1}}{3.9 \times 10^{-6}\,\text{atm}^{-1}} 5\,\text{K} = 232 \text{ atm},$$

$$(1.6)$$

the integration being carried out at constant volume.

What does this teach us? The coefficients β and k can be measured rather easily at constant pressure or temperature, respectively. A simple mathematical exercise (see Eqs. 1.3–1.5) then shows that it is not necessary to measure $(\partial p / \partial T)_V$ separately. The data for β and k can be used to predict this quantity on a firm basis. The following physical interpretation could be applied to this exercise. First heat the mercury from 25°C to 30°C at a constant pressure of 1 atm. The value of the coefficient of thermal expansion β allows us to calculate the volume change for this process. Now, at a constant temperature of 30°C, compress the mercury back to its original volume. The value of the isothermal compressibility k allows us to calculate the required pressure. This line of reasoning indicates that the desired result can be calculated from the given information, k and β.

This example illustrates how mathematics alone can guide the economic recording of data. The first and second laws of thermodynamics provide additional rules that allow some quantities to be predicted from others, and we thereby learn how to record thermodynamic data in a simple and useful form without duplication, ambiguity, or contradiction. We shall return to this in Chapter 7.

These last thoughts project thermodynamics as a macroscopic, exact science, not dependent on a molecular viewpoint for its validity. At the same time, it can be prosaic because it provides the framework for recording data but not for predicting it. On a second level, there is a place where molecular and chemical concepts can enter into thermodynamics. This is the subject known as statistical mechanics. Frequently, it is *not* possible to proceed rigorously to calculate thermodynamic properties on the basis of laws of molecular interaction, although we should like to do so. Nevertheless, molecular concepts do often account qualitatively for the trends in the data as well as the magnitude. They, thus, provide a means for understanding the physical phenomena as well as a guideline for correlating the thermodynamic data.

These considerations lead us to develop the following approach to thermodynamics. A coherent line of theoretical development is necessary in order to arrive at the concepts of entropy and the thermodynamic temperature scale on the basis of the second law of thermodynamics; this is carried out in Chapter 6. At the same time, a parallel development of the thermodynamic properties of pure substances is carried out in Chapters 3, 5, 9, and 11. This material is of vital necessity in the application of thermodynamics, which is, of course, the principal objective here.

Applications of the thermodynamics of pure substances include cycles, open systems, and phase transitions (Chapters 11–13). Cycles are directly related to the classical thermodynamic subject of the interconversion of mechanical and thermal energy. The Carnot cycle is indispensable to the theoretical development (see Section 6.2), and refrigeration cycles may still be of some practical interest to chemical engineers. Phase transitions in pure substances provide an introduction to phase equilibria.

Part B introduces the thermodynamics of multicomponent solutions with an eye on the eventual need of the chemical engineer to be able to treat chemical equilibria and phase equilibria.

Chapter 2

Concept of Temperature

Temperature is the measure of the intensity of a substance's internal thermal energy. Heat flows spontaneously from a hot object to a cold object, and two bodies that have come to thermal equilibrium possess the same temperature.

Let us consider at length how we should establish a temperature scale. Table 2.1 gives results of some temperature measurements made with four thermometers on systems at three different temperatures. The first requirement for a thermometer is, as we see, that it possess a property that responds to temperature. This thermometric property, as we call it, may be the resistance of a wire, the volume of a substance, the pressure exerted by a gas enclosed in a given volume, etc.

Criteria that we might consider in the selection of a temperature scale or a thermometer are as follows:

1. The thermometric property should have a linear dependence on temperature.
2. Ease of calibration.
3. The temperature scale should be independent of the substance (or means) used to measure it.
4. The range of applicability of the thermometer or the definition of temperature.

The Newman Lectures on Thermodynamics
John Newman and Vincent Battaglia
Copyright © 2019 Jenny Stanford Publishing Pte. Ltd.
ISBN 978-981-4774-26-0 (Hardcover), 978-1-315-10861-2 (eBook)
www.jennystanford.com

5. Reproducibility—is a measurement independent of extraneous factors?
6. Convenience.
7. Response time of thermometer.
8. Interference with system being measured.
9. Agreement with the thermodynamic temperature scale.

Table 2.1 Temperature measurements (made by Douglas N. Bennion at U.C.L.A., October 1, 1971).

System→	Water–Steam	Water–Ice	CO$_2$ Solid–Vapor (Acetone)
θ_1	25.6 cm	14.1 cm	—
θ_2	5.26 mV	−0.02 mV	−3.21 mV
θ_3	486 mV	344 mV	281 mV
θ_4	20.9 cm Hg	−4.8 cm Hg	−21.4 cm Hg

Note: θ_1 is the length of a mercury column in a glass capillary, the ordinary mercury thermometer. θ_2 is the potential of an iron–constantan thermocouple with the reference junction at the ice point. θ_3 is the potential across a broken light bulb carrying a constant current, that is, a resistance thermometer. θ_4 measures the pressure of a constant volume of air by means of an open-end mercury manometer (under conditions where the barometer stood at 75.4 cm Hg).

Criteria 4 and 5 are of utmost importance in the selection of a thermometer used to establish or define a primary temperature scale against which all secondary thermometers are calibrated. The constant-volume gas thermometer, θ_4 in Table 2.1, illustrates the problem of reproducibility. In order to obtain results that are reproducible from day to day, the height of the barometer must be added to the displacement of the mercury manometer used to sense the pressure in the thermometer. In order for this thermometer to respond to temperature in the same way in Berkeley or Denver as in Los Angeles, a further correction should be made for the local gravitational acceleration. Barometric pressure and location of the laboratory are examples of "extraneous factors" that we would not want to affect the result of a temperature measurement.

The mercury thermometer has a limited range of applicability compared to the other thermometers in Table 2.1; the mercury froze before the carbon dioxide point could be attained.

In order to define a temperature scale, every detail of the thermometer must be clearly specified. For a constant-volume gas thermometer, these details would include the following:

- The material of construction, which might be glass. The material should be inert and have a known coefficient of thermal expansion.
- The gas used, for example hydrogen or nitrogen. Hydrogen fluoride would be a particularly bad choice because of its corrosive and toxic character.
- The volume of gas.
- The mass of gas.
- The manometer fluid. This should be inert and have a low vapor pressure, so as not to contaminate the gas in the thermometer.

There now remain still several ways in which the temperature scale can be defined with this given primary thermometer. Let the thermometer scale t be defined to be a linear function of the thermometric property:

$$t = a + b\theta, \tag{2.1}$$

the thermometric property being the absolute pressure p in the case of the constant-volume gas thermometer defined above. When the constants a and b are selected so that $t = 0$ at the ice point and $t = 100$ at the steam point, this is a centigrade scale.

The definition of the temperature scale does not need to involve two fixed points. An alternative would be to use one,

$$T = A\theta, \tag{2.2}$$

or none:

$$T = \theta. \tag{2.3}$$

(And there is no fundamental reason why the chosen relation between the temperature scale T and the thermometric property must be linear.) The two-constant system, as in Eq. 2.1, was used for many years, but it has now been replaced by the one-constant system, as in Eq. 2.2. The single fixed point used is the triple point of water, which is assigned the value of $T = 273.16$ K. The triple point of water, where pure solid, liquid, and vapor coexist, is more reproducible than either the ice point or the steam point.

If one fixed point is better than two, one might conclude that no fixed point would be still better. However, the use of a fixed point and an adjustable constant, A in Eq. 2.2, allows the apparatus to be calibrated, which can compensate to a large extent for the unavoidable variations in the volume and mass of gas used in the construction of the thermometer.

One additional modification is involved in the definition of the currently used primary temperature scale. The above definition with the constant-volume gas thermometer shows a weak dependence on the mass m of gas used as well as a variation if, say, nitrogen is substituted for hydrogen. Let T_m be the temperature scale defined for a mass m of gas:

$$T_m = A_m \theta, \tag{2.4}$$

A_m being chosen such that $T_m = 273.16°$ at the triple point of water. Then, the "ideal-gas" temperature scale is that reached in the limit of a very dilute gas:

$$T = \lim_{m \to 0} T_m. \tag{2.5}$$

This is a so-called "absolute" temperature scale and happens to coincide with the Kelvin or thermodynamic temperature scale, developed in Section 6.3. The symbol K will be used to designate kelvin on this scale. The scale defined by Eq. 2.5 is found, by experiment, to be independent of which particular gas (H_2, N_2, O_2, H_2O, Ar, etc.) is used in the constant-volume thermometer.

The ideal-gas temperature scale is applicable over a wide range of temperature. It can be extended on the low end by means of the thermodynamic temperature scale and on the high end by means of Planck's law of radiation of black bodies.

The Celsius temperature scale, whose unit is denoted by °C, is defined by subtracting 273.15° from the Kelvin temperature scale. The triple point of water thus lies at exactly 0.01°C. The ice point is at approximately 0°C, and the steam point is at approximately 100°C. The Rankine scale (denoted by °R) is an absolute temperature scale whose degree is 1.8 times smaller than the Kelvin degree. It is thus defined by Eqs. 2.4 and 2.5 but with A_m chosen such that $T_m = 491.688°R$ at the triple point of water. The Fahrenheit temperature scale, whose unit is denoted by °F, is defined by subtracting 459.67° from the Rankine temperature scale.

Let us return to the criteria at the beginning of the chapter. The gas thermometer is not very convenient, has a long response time, and interferes considerably with the system being measured. And it is completely unfeasible to carry out the extrapolation to the ideal-gas scale (see Eqs. 2.4 and 2.5) every time you want to measure the temperature of a child with a fever. Consequently, it is expedient to define a practical, secondary temperature scale. Criterion 1 has meaning only for such secondary temperature scales. This was the desire for the thermometric property to have a linear dependence on temperature.

Problems

2.1 Discuss the advantages and disadvantages of the several thermometers in terms of such factors as convenience, speed of response, influence on the system being measured, accuracy, reproducibility, range of temperature covered, calibration, usefulness as a primary standard, and merit in establishing a temperature scale.

2.2 For each thermometer, let a temperature scale t_i be defined to be a linear function of the thermometric property:
$$t_i = a_i + b_i \theta_i.$$
Take the ice–water point to be $0°$ and the water–steam point to be $100°$, and use these fixed points to determine the constants a_i and b_i. (These are essentially centigrade temperature scales, not the Celsius temperature scale.) For the several thermometers, now determine t_i at the CO_2 solid–vapor point.

2.3 Discuss the relative merits of using zero, one, two, or more fixed points or calibration points to establish a temperature scale from a thermometric property.

2.4 What correction would be necessary in order to have the constant-volume gas thermometer give results that are independent of the day on which the measurement was made? What additional correction would be necessary in order for the device to give the same results in Berkeley or Denver as in Los Angeles?

2.5 Should one be concerned that some negative temperatures are recorded in Table 2.1?

Chapter 3

PVT Properties of Gases

Pressure is an important property of a substance and is subject to direct measurement as a function of the volume and temperature. The difference between a gas and a liquid is largely one of compressibility. Furthermore, computation of amounts of work, which enter into our energy balances, involves pressure and volume. The reversible work done by a substance is given by

$$dW_{\text{rev}} = pdV \qquad (3.1)$$

(for systems that do not do electrical work or surface work).

In this chapter, we develop the PVT relationships for pure substances. In later chapters, we will see how to make the fullest use of these results in the calculation of other thermodynamic properties of substances. This is partly because the PVT relationships can appeal to our physical senses and the measurements are more direct than for other thermodynamic properties.

3.1 Ideal Gases

We imagine that a suitable temperature scale θ has been defined. This might be, for example, the mercury-thermometer centigrade scale.

Boyle's law says that for a given amount of a gas at low pressures, the pressure is inversely proportional to the volume at constant temperature. In other words,

The Newman Lectures on Thermodynamics
John Newman and Vincent Battaglia
Copyright © 2019 Jenny Stanford Publishing Pte. Ltd.
ISBN 978-981-4774-26-0 (Hardcover), 978-1-315-10861-2 (eBook)
www.jennystanford.com

$$pV = F(\theta), \tag{3.2}$$

where F depends on the mass and nature of the gas as well as the temperature.

Charles's law says that for a given change in temperature from θ_1 to θ_2 at constant pressure, the relative volume change $\Delta V/V_1$ is the same for all gases. Gay-Lussac's law says practically the same thing; it says that the coefficient of thermal expansion

$$\beta = \frac{1}{V}\left(\frac{\partial V}{\partial \theta}\right)_p \tag{3.3}$$

is the same for all gases. Again, these laws apply at low pressures.

From these laws we can derive the ideal-gas law—the relationship between pressure, temperature, and volume, which applies to all gases in the limit of low pressures. This law provides a first approximation to the properties of gases and also allows us to establish the ideal-gas temperature scale.

Since the volume V is an extensive property, the function F in Eq. 3.2 must be proportional to the mass m of gas. The laws of Charles and Gay-Lussac tell us further that F is proportional to a universal function of temperature, that is, a function that is the same for all gases. We shall call this function $T(\theta)$. Consequently, Eq. 3.2 can now be expressed as

$$pV = mR_i\, T(\theta), \tag{3.4}$$

where R_i is a constant that can depend on the nature of the gas but is independent of both the mass m and the temperature θ.

The function T is now taken to define the ideal-gas temperature scale. The thermometric property is the quantity pV/m in the limit as the pressure approaches zero. The temperature scale is then normalized to the value 273.16 K at the triple point of water (denoted by the subscript t):

$$T = 273.16\frac{\lim\limits_{p\to 0}(pV/m)}{\lim\limits_{p\to 0}(pV/m)_t}. \tag{3.5}$$

To the extent that Eq. 3.4 represents the behavior of real gases as the pressure approaches zero, this definition of the ideal-gas temperature scale is complete and independent of the gas used as the thermometric substance.

The readers can also perceive that this scale is identical to that defined in Eqs. 2.4 and 2.5 on the basis of the constant-volume gas thermometer. The volume of a constant-pressure gas thermometer can also be used to define the ideal-gas temperature scale.

In Chapter 9, we show that the laws of Charles and Gay-Lussac can be derived from Boyle's law, that is, that $T(\theta)$ being independent of the nature of the gas follows directly from Eq. 3.2. It is shown that T is identical to the thermodynamic temperature scale.

Next, it is customary to use the properties of dilute gases, Eq. 3.4, to define the mole.[1] The constant R_i is written as

$$R_i = R/M_i, \tag{3.6}$$

where R is the universal gas constant and M_i is the molar mass of the gas. The standard now used for defining molar masses is that the isotope carbon 12 be given the value of exactly 12 g/mol. In the past, the standard had been that the naturally occurring isotopic mixture of O_2 should have the value of exactly 32. However, the carbon standard should be more reproducible. The readers might notice that the properties of gases can no longer be used by themselves to establish the molar masses of gases since carbon does not form a gas in any practical manner.

The molar masses of a number of common gases are given in Table 3.1. The universal gas constant R and other universal constants are given in the notation at the end of any chapter where they are used. The pound mole, abbreviated lb-mol, is 453.5924 times larger than the mole defined above.

The number of moles n is given as m/M_i. We use the tilde to denote quantities on a molar basis and the circumflex to denote quantities per unit mass. Thus, the extensive property, volume, can be written as

$$V = n\tilde{V} = m\hat{V}. \tag{3.7}$$

The density ρ is the reciprocal of $\hat{V}$.

The ideal-gas law 3.4 can now be written in the equivalent forms

$$\left. \begin{aligned} pV &= nRT \\ p\tilde{V} &= RT \\ p &= \rho RT/M_i \end{aligned} \right\} \tag{3.8}$$

---

[1]Abbreviated "mol" in the *Système International* (SI) units

The ideal-gas law is in harmony with molecular concepts and can be derived in statistical mechanics. The molecules have randomly distributed velocities. The collisions of these molecules with the walls make the dominant contribution to the pressure, for dilute gases (see Problem 3.3). At higher pressures, intermolecular forces cause deviations from the ideal-gas law.

3.2 A Real Substance

Figure 3.1 depicts the PVT relationship for a real substance, water. The ideal-gas law 3.8 becomes a better approximation for large values of $\hat{V}$, that is, toward the right on Fig. 3.1.

Table 3.1 Critical properties and molar masses of several substances.

| | M_l | T_c | p_c | Z_c |
	g/mol	K	atm	
helium (He)	4.0026	5.25	2.26	0.300
neon (Ne)	20.183	44.45	26.9	0.305
argon (Ar)	39.948	150.85	48	0.292
krypton (Kr)	83.80	209.35	54.3	0.290
xenon (Xe)	131.30	289.75	58	0.288
radon (Rn)	222.00	377.19	62	—
methane (CH_4)	16.04	191.05	45.8	0.289
tetrafluoromethane (CF_4)	88.00	227.45	41.4	—
carbon tetrachloride (CCl_4)	153.82	556.25	45	0.272
hydrogen (H_2)	2.0159	33.25	12.8	0.304
deuterium (D_2)	4.032	38.35	16.4	—
nitrogen (N_2)	28.0134	126.26	33.54	0.292
oxygen (O_2)	31.9988	154.75	50.1	0.292
fluorine (F_2)	37.996	144.15	55	—
chlorine (Cl_2)	70.906	417.15	76.1	0.276
bromine (Br_2)	—	575.15	—	—
iodine (I_2)	253.809	785.15	116	—
carbon dioxide (CO_2)	44.01	304.15	72.9	0.276

	M_i	T_c	p_c	Z_c
	g/mol	K	atm	
acetylene (HCCH)	26.04	308.65	61.6	0.274
hydrazine (NH$_2$NH$_2$)	32.05	653.15	145	—
sulfur (S)	—	1313.15	—	—
benzene (C$_6$H$_6$)	78.11	562.05	48.6	0.274
cyclohexane (C$_6$H$_{12}$)	84.16	553.55	40	0.271
carbon monoxide (CO)	28.01	133.15	34.5	0.294
hydrogen fluoride (HF)	20.01	461.15	64	—
hydrogen chloride (HCl)	36.46	324.55	82.1	0.266
hydrogen bromide (HBr)	80.92	363.15	84.5	—
hydrogen iodide (HI)	127.91	423.15	81.9	—
nitric oxide (NO)	30.01	180.15	64	0.25
nitrogen dioxide (NO$_2$)	46.01	430.95	100	—
nitrous oxide (N$_2$O)	44.01	309.65	71.7	0.271
ammonia (NH$_3$)	17.03	405.65	112.5	0.242
water (H$_2$O)	18.01534	647.25	218.3	0.250
hydrogen sulfide (H$_2$S)	34.08	373.55	88.9	0.284
sulfur dioxide (SO$_2$)	64.06	430.95	77.7	0.268
sulfur trioxide (SO$_3$)	80.06	491.45	83.6	0.263
carbon disulfide (CS$_2$)	76.14	552.15	78	—
ethane (C$_2$H$_6$)	30.07	305.35	48.2	0.284
ethylene (C$_2$H$_4$)	28.05	283.05	50.5	0.268
propane (C$_3$H$_8$)	44.09	369.95	42	0.276
1,3-butadiene (C$_4$H$_6$)	54.09	425.15	42.7	0.270
n-butane (C$_4$H$_{10}$)	58.12	425.15	37.5	0.274
n-pentane (C$_5$H$_{12}$)	72.15	469.75	33.3	0.268
n-hexane (C$_6$H$_{14}$)	86.18	507.35	29.9	0.264
styrene (C$_8$H$_8$)	104.14	647.55	39.4	—
p-xylene (C$_8$H$_{10}$)	106.16	618.15	33.9	—
n-octane (C$_8$H$_{18}$)	114.23	569.15	24.8	0.258
2,2,3-trimethyl-pentane (C$_8$H$_{18}$)	114.23	567.15	28.2	—
n-nonane (C$_9$H$_{20}$)	128.26	594.15	22.5	—

(Continued)

Table 3.1 (*Continued*)

	M_i	T_c	p_c	Z_c
	g/mol	K	atm	
naphthalene ($C_{10}H_8$)	128.18	747.95	40.6	—
n-decane ($C_{10}H_{22}$)	142.29	617.55	20.8	—
biphenyl ($C_{12}H_{10}$)	154.20	768.15	31.8	—
n-dodecane ($C_{12}H_{26}$)	170.34	659.15	17.9	—
monochloro-methane (CH_3Cl)	50.49	416.95	65.9	0.276
methylene chloride (CH_2Cl_2)	84.93	510.15	60	—
chloroform ($CHCl_3$)	119.38	536.15	54	0.283
monochlorotrifluoro-methane ($CClF_3$)	104.46	302	38.2	—
chlorodifluoromethane ($CHClF_2$, Freon 22)	86.48	369.16	49.12	0.267
trichlorotrifluoroethane (CCl_2FCClF_2, Freon 113)	187.38	487.38	33.7	0.274
n-heptane (C_7H_{16})	100.21	540.25	27	0.260
perfluoro-*n*-heptane (C_7F_{16})	388.7	474.85	16	—
1,1-dichloroethane ($C_2H_4Cl_2$)	98.96	522.95	50	—
methanol (CH_3OH)	32.04	513.15	78.5	0.220
acetonitrile (CH_3CN)	41.05	547.85	47.7	0.181
methyl mercaptan (CH_3SH)	48.11	469.95	71.4	0.276
nitromethane (CH_3NO_2)	61.04	587.95	62.3	—
methyl amine (CH_3NH_2)	31.06	430.05	40.2	—
acetic acid (CH_3COOH)	60.05	594.75	57.1	0.200
phenol (C_6H_5OH)	94.11	694.25	60.5	—
aniline ($C_6H_5NH_2$)	93.13	698.75	52.3	—
toluene ($C_6H_5CH_3$)	92.13	593.95	41.6	0.27
p-cresol ($CH_3C_6H_4OH$)	108.14	704.55	50.8	—
ethanol (C_2H_5OH)	46.07	516.15	63	0.249
isopropyl alcohol (C_3H_7OH)	60.09	508.15	47	—
n-propyl alcohol (C_3H_7OH)	60.09	536.75	51	0.255
n-butyl alcohol (C_4H_9OH)	74.12	562.95	43.6	—
tert-butyl alcohol (($CH_3)_3COH$)	74.12	508.15	39.2	—

	M_i	T_c	p_c	Z_c
	g/mol	**K**	**atm**	
1-heptyl alcohol ($C_7H_{15}OH$)	116.20	638.45	29.4	—
1-octyl alcohol ($C_8H_{17}OH$)	130.23	658.65	26.5	—
diethyl ether (($C_2H_5)_2O$)	74.12	465.75	35.6	0.261
acetone (($CH_3)_2CO$)	58.08	508.65	47	0.237
ethyl methyl ketone ($CH_3C_2H_5CO$)	72.11	535.15	41	0.26
ethylene oxide (C_2H_4O)	44.05	468.95	71	0.25
propylene oxide (C_3H_6O)	58.08	482.15	48.6	—

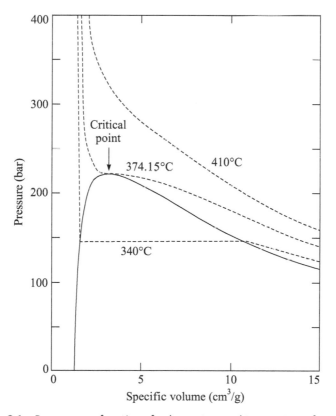

Figure 3.1 Pressure as a function of volume at several temperatures for water.

At certain temperatures, water in the form of liquid and vapor can coexist. This situation is represented by the horizontal line in

Fig. 3.1. The dashed, dome-shaped curve indicates the loci of saturated liquid and vapor. The maximum temperature at which distinct liquid and vapor phases can coexist is known as the *critical temperature* T_c, and has a corresponding *critical pressure* p_c and *critical volume* $\hat{V}_c$ or $\tilde{V}_c$. These critical properties are recorded in Table 3.1 for a number of pure substances.

The isotherm at the critical point has a zero slope in Fig 3.1. This is also an inflection point. Mathematically, these conditions are expressed as

$$\left(\frac{\partial p}{\partial V}\right)_T = 0 \ \text{ and } \ \left(\frac{\partial^2 p}{\partial V^2}\right)_T = 0 \qquad (3.9)$$

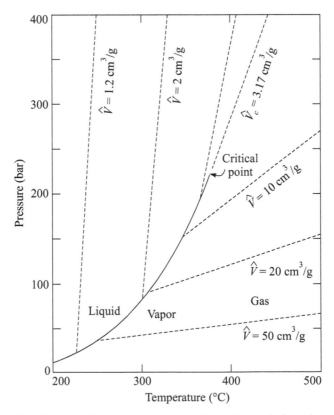

Figure 3.2 Pressure–temperature diagram for water, including the vapor–pressure curve.

At temperatures above the critical temperature, the substance is sometimes called a *gas*, the terms *liquid* and *vapor* being used below the critical temperature. Frequently, however, this distinction between a vapor and a gas is not maintained. As Fig. 3.1 shows, a liquid has low compressibility. On the other hand, the properties of liquid, vapor, and gas continuously blend into each other if one skirts around the solid curve in Fig. 3.1.

The data in Fig. 3.1 are replotted in Fig. 3.2 as pressure as a function of temperature with volume $\hat{V}$ as a parameter. The two-phase region in Fig. 3.1 now collapses into the vapor–pressure curve in Fig. 3.2. According to Eq. 1.5, the slope of a curve of constant volume (an *isochor*) is equal to β / k, the ratio of the coefficient of thermal expansion to the isothermal compressibility.

3.3 The van der Waals Equation

To account for deviations from the ideal-gas law, van der Waals proposed the relation

$$\left(p + \frac{a}{\tilde{V}^2} \right)(\tilde{V} - b) = RT , \tag{3.10}$$

where a and b are constants independent of p, $\tilde{V}$, and T but dependent on the nature of the substance. The constant b is supposed to represent the size of the molecules, so that $\tilde{V} - b$ can be regarded as the *free volume*. The term $-a / \tilde{V}^2$ represents the amount by which intermolecular attraction reduces the pressure, so that $p + a / \tilde{V}^2$ is the pressure due to the translational motion of the molecules.

Equation 3.10 is a relatively simple extension of the ideal-gas law, Eq. 3.8, since it contains only two adjustable constants a and b, and these have been endowed with some physical significance. The values of a and b are frequently selected so that Eq. 3.9 applies at the critical point. Combination of Eqs. 3.9 and 3.10 yields

$$a = \frac{27}{64} \frac{(RT_c)^2}{p_c} \quad \text{and} \quad b = \frac{RT_c}{8p_c} \tag{3.11}$$

so that values can be obtained from the critical temperature and pressure in Table 3.1. The critical volume is then predicted by the van der Waals equation to be

$$\tilde{V}_c = 3b = \frac{3RT_c}{8p_c}.$$

(3.12)

3.4 The Redlich–Kwong Equation

Redlich and Kwong [1] contended that a better representation for pure substances is given by the equation

$$p = \frac{RT}{\tilde{V} - b} - \frac{a}{T^{1/2}\tilde{V}(\tilde{V} + b)},$$

(3.13)

where a and b are constants (but different from those in the van der Waals equation). A fit of Eq. 3.13 to Eq. 3.9 at the critical point yields the values

$$a = 0.4275 \frac{R^2 T_c^{5/2}}{p_c} \quad \text{and} \quad b = 0.08664 \frac{RT_c}{p_c},$$

(3.14)

the critical volume then being predicted to be

$$\tilde{V}_c = \frac{RT_c}{3p_c}.$$

(3.15)

Equation 3.13 retains the advantage of having only two adjustable constants; more complicated equations of state will not be considered here.

3.5 Principle of Corresponding States

A study of data for several substances suggests that they are similar to each other. This concept finds quantitative expression in the principle of corresponding states: The deviation from the ideal-gas behavior, expressed in the *compressibility factor* $Z = p\tilde{V}/RT$, depends on the temperature and pressure in the reduced form T/T_c and p/p_c. The compressibility factor can be regarded as a dimensionless molar volume. In order to form a dimensionless temperature and a dimensionless pressure, one needs to select characteristic values for each substance. The critical point stands out as a natural choice for fluid properties. (The triple point, involving also the solid phase, is less appealing.)

Table 3.1 lists critical values for a number of substances. If the principle of corresponding states were rigorous, the values of Z_c in this table would be the same for all substances. These values are generally near 0.27. In contrast, the van der Waals equation predicts the value $Z_c = 0.375$, while the Redlich–Kwong equation predicts the value $Z_c = 0.333$ (see Eqs. 3.12 and 3.15).

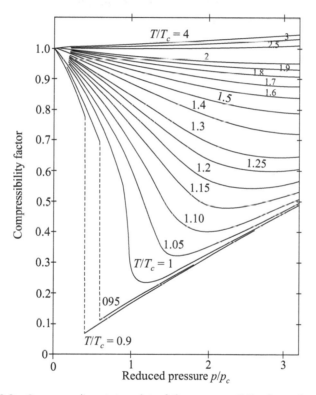

Figure 3.3 Corresponding-states plot of the compressibility factor for simple fluids, such as argon and neon. For correction for other normal fluids, based on the acentric factor, see Ref. [2].

Figure 3.3 is a corresponding-states plot of the compressibility factor Z for simple fluids. Accurate representation of the behavior of normal fluids requires the use of a correlating parameter in addition to reduced temperature and reduced pressure. Some authors have used the compressibility factor Z_c at the critical point (see Table 3.1);

others have used the *acentric factor* [2], based on the vapor pressure at a reduced temperature of 0.7. Figure 3.3 demonstrates explicitly how the product $p\tilde{V}$ approaches RT for small values of the pressure (see Eq. 3.2).

A corresponding-states plot of the vapor pressure can be found in Chapter 11.

3.6 Molecular Basis of the Principle of Corresponding States

Suppose that there is a class of substances for which the intermolecular force laws are similar. That is, there are symmetric molecules of mass m_l whose intermolecular potential energy function w is characterized by two parameters, an energy ε, and a distance σ (see Fig. 3.4), in such a way that w/ε as a function of r/σ is the same for all the substances in the class. We should expect the monatomic substances argon, neon, krypton, and xenon to form such a class, although methane might also belong to this class. Substances such as water and ammonia are less likely to fit into classes with other substances.

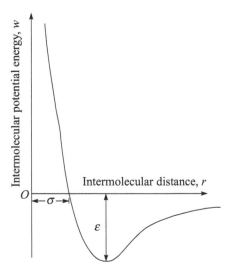

Figure 3.4 Intermolecular potential energy due to attraction and repulsion between molecules.

A representation of intermolecular potential energies, which is popular in theoretical studies, is the Lennard–Jones 6-12 potential:

$$w = 4\varepsilon \left[\left(\frac{\sigma}{r} \right)^{12} - \left(\frac{\sigma}{r} \right)^6 \right]. \tag{3.16}$$

Each substance that belongs to this class would then be characterized by a value of ε and a value of σ.

We wish to subject the PVT relationship for such fluids to dimensional analysis. The volume V will be taken to represent the total volume divided by the number of molecules. In addition to p, V, and T, we suppose that the mass m_i of the molecules, the Boltzmann constant k, and the parameters ε and σ of the intermolecular potential energy function are relevant. From these seven quantities, only three independent dimensionless groups can be formed, for example,

$$\frac{p\sigma^3}{\varepsilon}, \quad \frac{V}{\sigma^3}, \quad \text{and} \quad \frac{kT}{\varepsilon}.$$

These can be thought of as dimensionless or *reduced* pressure, volume, and temperature, respectively. Note that the mass m_i of a molecule does not enter into any of these dimensionless groups.

This exercise in dimensional analysis shows that the reduced volume V/σ^3 must be a dimensionless function of the reduced temperature kT/ε and the reduced pressure $p\sigma^3/\varepsilon$. The same functional relationship must apply to all substances belonging to the same class.

Furthermore, all substances in the same class will have the same values of critical properties when expressed as V_c/σ^3, $p_c\sigma^3/\varepsilon$, and kT_c/ε. Consequently, the molecular theory developed here is identical to the macroscopic principle of corresponding states discussed in Section 3.5.

The compressibility factor can be written as

$$Z = p\tilde{V}/RT = pV/kT \tag{3.17}$$

if V/k can be identified with $\tilde{V}/R$. Since V was taken here to be the volume per molecule, this means that the number of molecules per mole should be the same for all substances. The ratio $\tilde{V}/V = R/k$ has the value 6.0225×10^{23} mol^{-1} and is called Avogadro's or Loschmidt's number, denoted by L.

By recombining the dimensionless groups, we obtain

$$Z = Z\left(\frac{p\sigma^3}{\varepsilon}, \frac{kT}{\varepsilon}\right) \tag{3.18}$$

that is, the compressibility factor should be a universal function of the reduced pressure $p\sigma^3/\varepsilon$ and the reduced temperature kT/ε. In particular, Z is independent of the molecular mass m_i. Consequently, the product $p\tilde{V}$ should not depend explicitly on the molar mass of the substance.

3.7 The Virial Equation

For dilute gases, the compressibility factor can be expressed as a power series in the reciprocal of the molar volume:

$$Z = p\tilde{V}/RT = 1 + B/\tilde{V} + C/\tilde{V}^2 + D/\tilde{V}^3 + \cdots. \tag{3.19}$$

This reduces to the ideal-gas Eq. 3.8 in the limit of low pressures or large molar volumes. The higher-order terms in Eq. 3.19 account for the deviation from the ideal-gas behavior at higher pressures. The coefficients in these terms depend on temperature and the nature of the gas but are independent of $\tilde{V}$ or p.

One nice feature of the virial equation is that it represents a straightforward correction to the ideal-gas equation in the very region where that equation is valid, namely, at low pressures. It does not attempt to describe the behavior of a fluid near the critical point and at high densities, as might be expected of the van der Waals equation or the Redlich–Kwong equation. Consequently, the virial coefficients have a well-defined physical significance. They are not just fitting parameters.

For example, the second virial coefficient can be related to the intermolecular potential energy w according to the equation

$$B = -2\pi L \int_0^\infty [e^{-w/kT} - 1]r^2 dr. \tag{3.20}$$

B arises from collisions between molecules, two at a time, for which the interaction is described by w. Similarly, C arises from collisions involving three molecules simultaneously. The importance of binary collisions is proportional to $\tilde{V}^{-1}$, and the importance of ternary collisions is proportional to $\tilde{V}^{-2}$, as reflected in Eq. 3.19.

An alternative form of the virial equation is

$$Z = p\tilde{V}\,/\,RT = 1 + B'p + C'p^2 + D'p^3 + \cdots, \qquad (3.21)$$

where B', C', D',... again depend on temperature and the nature of the substance. This is entirely equivalent to Eq. 3.19, but now deviations from ideal-gas behavior are expressed by a power series in the pressure p. By a careful matching of the terms in the two equations (see Example 3.1), we can obtain the values of the coefficients in Eq. 3.21 from those in Eq. 3.19, with the result

$$B' = B/RT, \qquad C' = (C-B^2)/(RT)^2, \ldots\, . \qquad (3.22)$$

On the basis of the principle of corresponding states, it should be possible to relate the virial coefficients, in the reduced form, to the reduced temperature. Such a correlation is presented in Fig. 3.5. The data can be adequately represented by the equation [3]

$$B/\tilde{V}_c = 0.438 - 0.881(T_c/T) - 0.757(T_c/T)^2. \qquad (3.23)$$

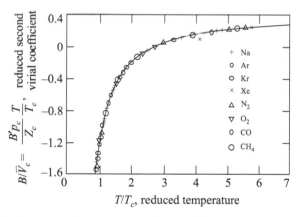

Figure 3.5 Corresponding-states correlation of the second virial coefficient [3].

Notice that $B'p_c$ should be the slope of the corresponding-states curve for the compressibility factor (see Fig. 3.3) at low pressures.

Figure 3.5 and Eq. 3.23 are particularly valuable since they describe accurately the behavior of real gases at low pressures. The critical properties for a number of substances are presented in

Table 3.1, and for these substances, we can now estimate the second virial coefficient. The behavior of gases is, for many purposes, adequately represented by the virial equations with terms beyond the second neglected, as discussed in the next section.

3.8   Truncated Virial Equation

For practical calculations, we shall often neglect terms beyond the second in the virial Eqs. 3.19 and 3.21. Thus, we write

$$p\widetilde{V}/RT = 1 + B/\widetilde{V} \qquad (3.24)$$

or

$$p\widetilde{V}/RT = 1 + B'/p. \qquad (3.25)$$

One can immediately obtain an idea of the adequacy of Eq. 3.25 from Fig. 3.3. When the curves in that figure deviate significantly from a straight line, Eq. 3.25 becomes an inadequate approximation. This occurs at different reduced pressures for different reduced temperatures.

One should notice that Eqs. 3.24 and 3.25 are not equivalent, and the reader might wonder which is more accurate. The answer depends on the value of the reduced temperature. To answer this question, let us use both the van der Waals equation and the Redlich–Kwong equation. We do this not because these equations themselves are accurate but because we should like to illustrate the relationship among these several representations of PVT properties and we should like to see how accurately the van der Waals equation and the Redlich–Kwong equation can predict the second virial coefficient.

By expanding the van der Waals Eq. 3.10 in a virial form, we obtain for the second and third virial coefficients (see Example 3.2)

$$B = b - a/RT \quad \text{and} \quad C = b^2 \qquad (3.26)$$

or

$$\frac{B}{\widetilde{V}_c} = \frac{1}{3} - \frac{9T_c}{8T} \quad \text{and} \quad \frac{C}{\widetilde{V}_c^2} = \frac{1}{9}. \qquad (3.27)$$

The last equations are obtained by means of Eqs. 3.11 and 3.12.

By expanding the Redlich–Kwong Eq. 3.13 in a virial form, we obtain for the second and third virial coefficients

$$B = b - \frac{a}{RT^{3/2}} \quad \text{and} \quad C = b^2 + \frac{ab}{RT^{3/2}} \tag{3.28}$$

or

$$\frac{B}{\tilde{V}_c} = 0.2599 - \frac{1.28244}{(T/T_c)^{3/2}} \quad \text{and} \quad \frac{C}{\tilde{V}_c^2} = 0.06756 + \frac{0.3338}{(T/T_c)^{3/2}}. \tag{3.29}$$

The last equations are obtained by means of Eqs. 3.14 and 3.15.

From these results, B' and C' can be obtained by means of Eq. 3.22.

The first neglected term in Eq. 3.25 is $C'p^2$, and the first neglected term in Eq. 3.24 is $C/\tilde{V}^2$, which can be approximated by $p^2C/(RT)^2$.

The value of the reduced pressure p/p_c at which the first neglected term is 0.01 (representing about 1% of the compressibility factor) can thus be expressed as $0.1/p_c\sqrt{|C'|}$ for Eq. 3.25 and as $0.1\,RT/p_c\sqrt{C}$ for Eq. 3.24. Table 3.2 gives values of these quantities as functions of the reduced temperature T/T_c. These results are based on the van der Waals equation, the Redlich–Kwong equation, and experimental virial coefficients for methane.

We conclude that Eq. 3.25 is a better approximation than Eq. 3.24 for reduced temperatures above about 1.2, and vice versa. Either equation should be accurate up to saturation for reduced temperatures less than 0.8. Since critical pressures are on the order of 40 atm, the truncated equations should generally be accurate up to at least 10 atm, which frequently covers the range of interest.

Table 3.2 Region of accuracy of truncated virial equations, and comparison of van der Waals and Redlich–Kwong equations with experimental data for methane.

	van der Waals		Redlich–Kwong			Methane			
T/T_c	$\dfrac{0.1}{p_c\sqrt{|C'|}}$	$\dfrac{0.1RT}{p_c\sqrt{C}}$	$\dfrac{0.1}{p_c\sqrt{|C'|}}$	$\dfrac{0.1RT}{p_c\sqrt{C}}$	T (°C)	B $\frac{cm^3}{mol}$	C $\frac{cm^6}{mol^2}$	$\dfrac{0.1}{p_c\sqrt{|C'|}}$	$\dfrac{0.1RT}{p_c\sqrt{C}}$
0.7	0.152	0.560	0.120	0.263	—	—	—	—	—
0.8	0.209	0.640	0.178	0.329	—	—	—	—	—
0.9	0.281	0.720	0.259	0.399	—	—	—	—	—
1.0	0.371	0.800	0.374	0.474	—	—	—	—	—

(Continued)

Table 3.2 (*Continued*)

T/T_c	van der Waals		Redlich–Kwong		Methane				
	$\dfrac{0.1}{p_c\sqrt{\lvert C'\rvert}}$	$\dfrac{0.1RT}{p_c\sqrt{C}}$	$\dfrac{0.1}{p_c\sqrt{\lvert C'\rvert}}$	$\dfrac{0.1RT}{p_c\sqrt{C}}$	T (°C)	B $\dfrac{cm^3}{mol}$	C $\dfrac{cm^6}{mol^2}$	$\dfrac{0.1}{p_c\sqrt{\lvert C'\rvert}}$	$\dfrac{0.1RT}{p_c\sqrt{C}}$
1.1	0.486	0.880	0.543	0.553	—	—	—	—	—
1.2	0.635	0.960	0.824	0.635	—	—	—	—	—
1.3	0.836	1.040	1.434	0.721	—	—	—	—	—
1.4	1.126	1.120	6.392	0.810	—	—	—	—	—
1.430	1.241	1.144	2.866	0.837	0	−53.35	2620	3.227	0.956
1.692	13.311	1.354	1.498	1.085	50	−34.23	2150	1.845	1.249
1.954	2.278	1.563	1.536	1.346	100	−21.00	1834	1.791	1.561
2.215	2.080	1.772	1.705	1.618	150	−11.40	1640	1.951	1.872
2.477	2.126	1.982	1.930	1.900	200	−4.16	1514	2.191	2.179
2.739	2.253	2.191	2.192	2.188	250	1.49	1420	2.489	2.487
3.001	2.420	2.401	2.482	2.481	300	5.98	1360	2.822	2.784
3.263	2.612	2.610	2.798	2.778	350	9.66	1330	3.176	3.061

Note: Here $0.1/p_c\sqrt{\lvert C'\rvert}$ is an estimate of the value of p/p_c at which Eq. 3.25 is in error by 1%, and $0.1\,RT/p_c\sqrt{C}$ is an estimate of the value of p/p_c at which Eq. 3.24 is in error by 1%.

The introduction to this book shows the power of the truncated virial equation. This is more fully appreciated in the treatment of chemical equilibria in Chapter 19.

Example 3.1 Express the virial Eq. 3.19 in the form of the virial Eq. 3.21 and thereby obtain the relationships 3.22 between the two sets of virial coefficients.

Solution: The virial equations are expansions for large volumes $\tilde{V}$ or small pressures p. Equation 3.19 yields a first approximation for the volume in terms of the pressure

$$\tilde{V} = \frac{RT}{p} + O(1),$$

the order of the first neglected term being denoted by the symbol O. Substitution of this result back into Eq. 3.19 yields a second approximation to the volume:

$$\widetilde{V} = \frac{RT}{p}\left[1 + \frac{Bp}{RT}\right] + O\,(p).$$

This process could be continued to obtain successively better approximations to $\widetilde{V}$.

At any level of approximation, the compressibility factor can be expressed as a power series in pressure. Substitution of the above result into Eq. 3.19 yields

$$\frac{p\widetilde{V}}{RT} = 1 + \frac{B}{\dfrac{RT}{p}\left[1 + \dfrac{Bp}{RT} + O(p^2)\right]} + \frac{C}{\left(\dfrac{RT}{p}\right)^2 [1 + O(p)]^2} + O(p^3)$$

$$= 1 + \frac{Bp}{RT}\left[1\;\frac{Bp}{RT} + O(p^2)\right] + \frac{Cp^2}{(RT)^2} + O(p^3)$$

$$= 1 + \frac{Bp}{RT} + \frac{C - B^2}{(RT)^2}p^2 + O(p^3).$$

Comparison of this form with Eq. 3.21 gives the relationships 3.22.

Example 3.2 Expand the van der Waals equation in a virial form and obtain expressions for the second and third virial coefficients for a van der Waals fluid.

Solution: The van der Waals Eq. 3.10 can be written exactly as

$$\frac{p\widetilde{V}}{RT} = 1 + \frac{pb}{RT} - \frac{a}{RT\widetilde{V}} + \frac{ab}{RT\widetilde{V}^2}\ .$$

As the first approximation to the pressure, we thereby obtain

$$p = \frac{RT}{\widetilde{V}} + O(\widetilde{V}^{-2}).$$

Substitution of this result into the first equation gives a second approximation to the pressure,

$$p = \frac{RT}{\widetilde{V}} + \frac{bRT - a}{\widetilde{V}^2} + O(\widetilde{V}^{-3}),$$

and substitution of this second approximation into the first equation yields

$$\frac{p\widetilde{V}}{RT} = 1 + \frac{b - a/RT}{\widetilde{V}} + \frac{b^2}{\widetilde{V}^2} + O(\widetilde{V}^{-3}).$$

Equations 3.26 follow from a comparison of this result with the virial Eq. 3.19.

Problems

3.1 Compare values of $B'p_c$ in Fig. 3.5 with the slopes at zero pressure in Fig. 3.3.

3.2 For a lecture-demonstration experiment, it is desired to construct a sealed glass vial that contains a pure substance, which can be made to pass through the critical point by heating the vial in a person's hand. Thus, at room temperature, the vial should contain a liquid and its vapor.

    a. From the list of critical properties, select a suitable substance to be sealed within the vial.

    b. What magnitude of pressures must the vial withstand?

    c. For a vial containing 100 cm^3, how much of the substance should be enclosed in the vial?

    d. Describe the changes within the vial as it is heated if it contains an amount of substance that is less than, equal to, and greater than that calculated in Problem 3.2c.

3.3 The Maxwell–Boltzmann distribution applies to the velocities of the molecules of a substance:

$$N_v(\underline{v}) = A \exp\left\{-\frac{m_i v^2}{2kT}\right\},$$

where m_i is the mass of the molecule, k is the Boltzmann constant $(m_i/k = M_i/R)$, and A is a constant independent of the velocity.

If momentum changes of molecules at a wall are the sole contribution to the pressure, show that the Maxwell–Boltzmann distribution of velocities implies the ideal-gas law 3.8.

3.4 The *Boyle temperature* is that for which

$$\left(\frac{\partial p\tilde{V}}{\partial p}\right)_T = 0$$

in the limit of zero pressure, and thus Boyle's law is accurate to moderately high pressures.

    a. Show that the second virial coefficient is zero at the Boyle temperature.

    b. Pick out the isotherm for the Boyle temperature on the compressibility-factor chart 3.3.

    c. Show that the Boyle temperature occurs at $T = a/Rb$ or at a reduced temperature of 3.375 for a van der Waals fluid.

    d. Show that the Boyle temperature occurs at $T = (a/Rb)^{2/3}$ or at a reduced temperature of 2.898 for a Redlich–Kwong fluid.

    e. Show that the Boyle temperature occurs at a reduced temperature of 2.661 for a fluid obeying the corresponding-states Eq. 3.23.

3.5 Investigate the accuracy of the van der Waals equation and the Redlich–Kwong equation by calculating second virial coefficients from Eqs. 3.27 and 3.29 and comparing these results with the value given by the corresponding-states Eq. 3.23. Do this at reduced temperatures of 0.5, 1, 1.5, 2, 3, 4, 5, and 6.

Notation

a	van der Waals constant, atm-cm^6/mol^2, or Redlich–Kwong constant, K$^{1/2}$-atm-cm^6/mol^2
b	van der Waals constant or Redlich–Kwong constant, cm^3/mol
B	second virial coefficient, cm^3/mol
B'	second virial coefficient in pressure expansion, atm^{-1}
C, D	third and fourth virial coefficients
C', D'	third and fourth virial coefficients in pressure expansion
$k = R/L$	Boltzmann constant, 1.38×10^{-23} J/K
L	Avogadro's (or Loschmidt's) number, 6.0225×10^{23} mol^{-1}
m	mass, g
m_i	molecular mass, g
M_i	molar mass of species i, g/mol
n	number of moles
P	pressure, atm or bar
r	intermolecular distance, cm
R	universal gas constant, 8.3143 J/mol-K or 82.06 atm-cm^3/mol-K

$R_i = R/M_i$ species gas constant, J/g-K

T absolute temperature, K

V volume, cm^3

$\tilde{V}$ molar volume, cm^3/mol

$\hat{V} = 1/\rho$ volume per unit mass, cm^3/g

w intermolecular potential energy, J

W_{rev} reversible work, J

$Z = p\tilde{V}/RT$ compressibility factor

β coefficient of thermal expansion, K^{-1}

ε energy at minimum in intermolecular potential energy, J

θ arbitrary measure of temperature

ρ density, g/cm^3

σ distance characteristic of intermolecular-potential-energy function, cm

Subscript

c critical point

References

1. Otto Redlich and J. N. S. Kwong, "On the thermodynamics of solutions. V. An equation of state. Fugacities of gaseous solutions," *Chemical Reviews*, **44**, 233–244, 1949.

2. Gilbert Newton Lewis and Merle Randall, revised by Kenneth S. Pitzer and Leo Brewer, *Thermodynamics* (New York: McGraw-Hill Book Company, Inc., 1961), pp. 606–609.

3. E. A. Guggenheim, *Thermodynamics* (Amsterdam: North Holland Publishing Company, 1959), p. 168.

Chapter 4

Conservation of Energy

The total energy of an isolated system does not change with time. The total energy of the system includes kinetic energy, potential energy, and internal energy, as we shall develop later in this chapter.

For a system which is not isolated, we need to be able to identify energy in transit. For a given problem or a given process, the material on which we focus our attention is called the *system*, or the *thermodynamic system* (see Fig. 4.1). The material with which the system interacts is called the *surroundings*, and the system and the surroundings together are called the *universe*. The term universe is not used in any cosmic sense; it is merely the least amount of relevant matter that forms an isolated system.

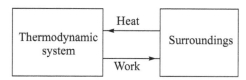

Figure 4.1 The thermodynamic system together with its immediate surroundings constitute the *universe*, an isolated system whose total energy is conserved. Work and heat are forms of energy in transit between the system and its surroundings.

The thermodynamic system can be either *open* or *closed* depending on whether or not any matter is transferred between the

The Newman Lectures on Thermodynamics
John Newman and Vincent Battaglia
Copyright © 2019 Jenny Stanford Publishing Pte. Ltd.
ISBN 978-981-4774-26-0 (Hardcover), 978-1-315-10861-2 (eBook)
www.jennystanford.com

system and its surroundings. We shall at first consider only closed systems. Since the total energy of an isolated system is conserved, the energy change of the system and that of the surroundings during a process must be equal and opposite, and some energy must have been transferred between them. We can distinguish two principal kinds of energy in transit between the system and its surroundings:

1. The work W (conventionally taken to be that done by the system) is the product of a force and the distance over which the force operates.

2. The heat Q (conventionally taken to be that added to the system) can be said to be that remaining energy in transit that is necessary in order to ensure that energy is conserved. It can be measured by its relationship to energy and work.

Let E_{tot} represent the total energy of the system, as distinguished from its surroundings. Then, for a process, the law of conservation of energy can be written as

$$\Delta E_{tot} = E_{tot}(\text{final}) - E_{tot}(\text{initial}) = Q - W. \tag{4.1}$$

Perhaps we should give some examples in order to clarify these concepts. In these examples, focus your attention on the fact that the measurement of work is the principal means of assessing both heat and total energy.

Let the system be a cannon ball of mass m. Work can be done on the cannon ball by raising it against the force of gravity. If this is done slowly, so that inertial effects are not important, then the force exerted by the cannon ball on its surroundings is

$$F = mg, \tag{4.2}$$

where g is the gravitational acceleration (see Fig. 4.2). If the cannon ball is raised in this manner from a vertical position h_1 to a vertical position h_2, the amount of work done *by* the cannon ball is

$$W = -mg(h_2 - h_1), \tag{4.3}$$

a negative quantity since the force and the distance moved are in opposite directions.

If no heat has been transferred, then the total energy of the system (the cannon ball) must have been changed by the amount of work done, in accordance with Eq. 4.1. (We can ensure that no heat is transferred by coating the cannon ball with a thermally insulating film or by maintaining the surroundings at the same temperature as the system.) The work was an interaction with the surroundings, that is, the mechanism that raised the cannon ball.

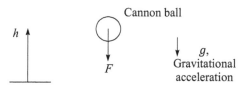

Figure 4.2 A cannon ball as a gravitational system. In subsequent chapters, this can be regarded as an example of a *reversible work source*.

In the absence of gravitational effects, the application of a force on the cannon ball will cause it to accelerate. We could imagine, for example, that the cannon ball in Fig. 4.2 is accelerated horizontally, in a direction perpendicular to the gravitational acceleration. Newton's second law of motion then expresses the force (exerted by the cannon ball on its surroundings) as

$$F = -\frac{d(mv)}{dt},$$ (4.4)

where v is the velocity of the cannon ball. The distance through which this force acts in a time element dt is $dl = v dt$. Hence, the work done by the cannon ball when it is accelerated from the velocity v_1 to the velocity v_2 is

$$W = \int_{t_1}^{t_2} F \cdot dl = -\int_{t_1}^{t_2} \frac{d(mv)}{dt} \cdot v dt$$

$$= -\frac{1}{2}m\left(v_2^2 - v_1^2\right).$$ (4.5)

If no heat has been transferred, then the total energy of the system must, again, have been changed by the amount of work done, in accordance with Eq. 4.1.

From these two processes, we can identify two contributions to the total energy E_{tot}. These are the potential energy

$$E_p = mgh$$ (4.6)

and the kinetic energy

$$E_k = \frac{1}{2}mv^2.$$ (4.7)

These are both forms of mechanical energy, studied in the branch of physics known as mechanics. The basic method of measuring changes in potential energy and kinetic energy is to measure work and to apply Eq. 4.1. It is then found that these forms of energy can be related to the height and the velocity, and subsequent measurements need involve only these coordinates. Arbitrary constants could be added to the definitions 4.6 and 4.7 of the potential and kinetic energies without affecting the results of any measurements, since Eq. 4.1 involves only *changes* in the total energy.

When the cannon ball is allowed to fall freely, no work is involved. With no heat transfer, the total energy is constant. This means that potential energy is converted into kinetic energy. Now suppose that the cannon ball strikes a hard floor and bounces, coming to rest eventually after bouncing several times. If no work is transmitted to the floor and no heat is transferred, then there must be yet one more form of energy. This will be called *internal energy*, and given the special symbol U, since it depends on coordinates internal to the system and not on the external coordinates, height and velocity.

The internal energy U is of central interest in thermodynamics since it is basically different from the forms of energy that are studied in mechanics. Now the internal nature of matter becomes important. The principal internal coordinate on which the internal energy depends is the temperature (see Chapter 2). We are likely to think of the pressure or volume as the second internal coordinate. Only one of these two need be thought of as an independent variable, since the pressure and volume are related through the equation of state, discussed in Chapter 3. We shall concentrate our attention on systems that are adequately described by the temperature and the volume or pressure. However, for systems of several components, the composition must be specified; for systems with appreciable interfacial area, the surface tension and surface area will be important; for electrochemical systems, the electric charge and potential are appropriate; and in other cases, magnetization, electric polarization, extent of chemical reaction, etc. need to be mentioned.

We thus see that it is expedient to break the total energy E_{tot} down into the sum of three terms:

$$E_{tot} = E_p + E_k + U, \qquad (4.8)$$

where the potential energy E_p and the kinetic energy E_k are independent of the internal coordinates and the internal energy U is independent of the external coordinates h and v. Equation 4.1 then becomes

$$\Delta U + \Delta(mgh) + \Delta\left(\frac{1}{2}mv^2\right) = Q - W. \tag{4.9}$$

The primary measure of U is the work done. Figure 4.3 depicts a system for measuring the internal energy of water. The insulation is to ensure that the heat Q is zero, and the work is that done by passing an electric current I through a resistor, if the volume is held constant through the process (see Eq. 3.1). Here, the change in internal energy is

$$\Delta U = -W = I\Phi t, \tag{4.10}$$

where Φ is the applied potential and t is the time during which the current was allowed to flow. However, a differential quantity that is frequently measured and reported is the change in internal energy with temperature.

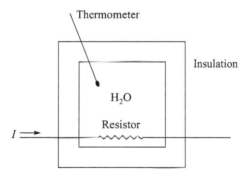

Figure 4.3 Constant-volume device for measuring changes in internal energy. The resistor is part of the thermodynamic system.

The *heat capacity* at constant volume is defined as

$$C_V = \left(\frac{\partial U}{\partial T}\right)_V. \tag{4.11}$$

Calorimetry is discussed further in the introduction to Chapter 14, in terms of complexities of multicomponent systems.

Next we remove the restriction of thermal insulation and introduce the concept of *heat*. Equation 4.9 is used to measure the heat Q after the dependence of U on internal coordinates such as temperature and pressure has been established through *adiabatic* or thermally insulated experiments. Work again becomes the basic measurement of Q since the other terms in Eq. 4.9 are established on this basis.

Equation 4.9 is the expression of conservation of energy or the *first law of thermodynamics*. We should notice that this equation says more than the law of conservation of energy as we first encountered it in mechanics. First, it recognizes that there is a second means besides work of exchanging energy between the system and the surroundings (see Fig. 4.1). This new form of energy in transit is called heat. Unlike work, it is not a product of a force and a distance (although, of course, it has the same dimensions). Second, Eq. 4.9 says that energy is conserved even though mechanical energy is not conserved. Two forms of thermal energy—internal energy and heat—are introduced, and mechanical energy can be converted into thermal energy, and vice versa. (The second law of thermodynamics places some restrictions on this interconversion of thermal and mechanical energy; see Section 6.4 and Chapter 8. Mechanical energy can always be converted completely into thermal energy, as depicted, for example, in Fig. 4.3.) On a microscopic level, the internal energy can be associated with the kinetic energy of molecules and the potential energy due to intermolecular forces and is thus not completely unrelated to mechanical energy. Statistical mechanics is the field that relates the properties of matter to the molecular picture.

It should be noticed that heat is quite distinct from internal energy. Although both are forms of thermal energy, the internal energy is a property of the thermodynamic system, and heat is energy in transit between the system and the surroundings (see Fig. 4.1). The first law of thermodynamics would have little to say if it could not state with certainty that the internal energy U is a property of the system, depending on the appropriate coordinates, such as temperature and pressure. Heat, then, is to be detected or measured by its interaction with the internal energy of matter. Figure 4.4 depicts a system in thermal contact with its surroundings. In the absence of work, either pressure–volume or electrical, the law of conservation of energy becomes

$$\Delta U = Q. \tag{4.12}$$

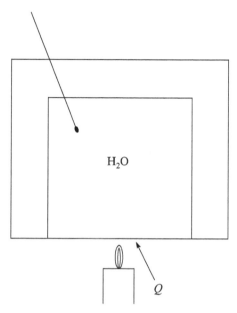

H₂O

Figure 4.4 Heating of water at constant volume. Here, no work is done on the water.

This explains the origin of the term *heat capacity* since in this case

$$\frac{dQ}{dT} = \left(\frac{\partial U}{\partial T}\right)_V = C_V. \tag{4.13}$$

The term is, however, something of a misnomer, since heat capacity is a property of the system, while heat is not. To emphasize that work and heat are not properties of the system, we shall occasionally put a slash through the *d* in the differential, thus: *đQ* and *đW*.

When our emphasis is on thermodynamics and the thermal properties of matter, we shall frequently express conservation of energy without the potential and kinetic energy terms. Then Eq. 4.9 becomes

$$\Delta U = Q - W \tag{4.14}$$

or in differential form

$$dU = đQ - đW. \tag{4.15}$$

Example 4.1 is designed to illustrate the relative importance of these energy terms.

4.1   Constant-Pressure Processes

The analysis of constant-pressure processes is of special practical importance because the atmosphere provides a constant-pressure environment. Figure 4.5 depicts a system within a cylinder and subjected to a constant pressure p by means of the piston. The work done by the system against this constant pressure will be

$$W = p\Delta V, \tag{4.16}$$

and in the absence of other forms of work, Eq. 4.9 becomes

$$\Delta U + \Delta(mgh) + \Delta\left(\frac{1}{2}mv^2\right) = Q - p\Delta V. \tag{4.17}$$

This equation can be simplified by introducing a new property of the system. The *enthalpy H* is defined as

$$H = U + pV. \tag{4.18}$$

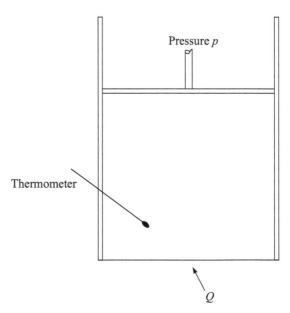

Figure 4.5   Heating a system at constant pressure.

For a constant-pressure process,

$$\Delta H = \Delta U + p\Delta V, \tag{4.19}$$

and conservation of energy for these special processes can be expressed as

$$\Delta H + \Delta(mgh) + \Delta\left(\frac{1}{2}mv^2\right) = Q. \tag{4.20}$$

In cases where the emphasis is on the thermal energy, we can write

$$\Delta H = Q, \tag{4.21}$$

neglecting the potential and kinetic energies. This result can be generalized somewhat by writing

$$\Delta H = Q - W', \tag{4.22}$$

where W' represents any work done by the system in addition to that required to change the volume against the constant pressure p. Thus, Fig. 4.6 depicts a constant-pressure system analogous to Fig. 4.3. These systems are adiabatic but can involve electrical work. In Figs. 4.4 and 4.5, the work, if any, is pressure–volume work.

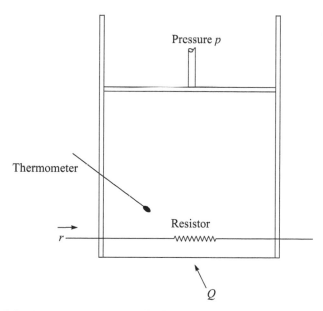

Figure 4.6 Apparatus to measure the heat capacity at constant pressure.

From Eq. 4.18, we see that the enthalpy is a function of the same variables as the internal energy. It can be regarded as a convenient, auxiliary property of matter that, for constant-pressure processes, accounts for any pressure–volume work and allows the researcher to concentrate his/her attention on thermal energy and other forms of work. These processes are so prevalent that the enthalpy is usually tabulated as a thermodynamic property of matter instead of the internal energy.

For processes with only pressure–volume work, Eq. 4.15 becomes

$$đQ = dU + pdV = dH - Vdp, \qquad (4.23)$$

and the heat capacity is defined as the heat added $đQ$ divided by the temperature change dT. For constant-volume processes, this led to the definition of the heat capacity at constant volume C_V as indicated in Eq. 4.13. The heat capacity is just as often measured with the constraint of a constant pressure, leading to the definition of a different heat capacity C_p:

$$C_p = \left(\frac{dQ}{dT}\right)_p = \left(\frac{\partial H}{\partial T}\right)_p, \qquad (4.24)$$

the last relationship following from Eq. 4.23.

We should like to be able to relate C_p to C_V. The differential of H can be written in general as

$$dH = \left(\frac{\partial H}{\partial T}\right)_p dT + \left(\frac{\partial H}{\partial p}\right)_T dp = C_p dT + \left(\frac{\partial H}{\partial p}\right)_T dp, \qquad (4.25)$$

and Eq. 4.23 becomes

$$đQ = C_p dT + \left[\left(\frac{\partial H}{\partial p}\right)_T - V\right] dp. \qquad (4.26)$$

Consequently, the heat capacity at constant volume is

$$C_V = C_p + \left[\left(\frac{\partial H}{\partial p}\right)_T - V\right]\left(\frac{\partial p}{\partial T}\right)_V. \qquad (4.27)$$

Not until Chapter 7 will we be able to show that, in general,

$$\left(\frac{\partial H}{\partial p}\right)_T - V = -T\left(\frac{\partial V}{\partial T}\right)_p. \qquad (4.28)$$

Let us mention in passing that, in measuring the heat capacity of a pure liquid, the constraint is frequently that the pressure is equal to the vapor pressure p_{sat}, and one is led to define a heat capacity

C_{sat}, for liquid maintained in equilibrium with its vapor. Equations 4.26 and 4.28 then yield

$$C_{sat} = C_p - T\left(\frac{\partial V}{\partial T}\right)_p \frac{dp_{sat}}{dT}. \tag{4.29}$$

Experimental values of C_{sat} are usually converted to values of C_p for the tabulation of data.

4.2 Steady Flow Processes

The equipment in Fig. 4.7 operates in a steady state, carrying out some process on the flowing streams. A pump forcing a material from a region of low pressure to a region of high pressure constitutes such a *steady flow process*. Under suitable operation, this class would also include heat exchangers (with two streams flowing in and two streams flowing out), a distillation column (which effects a separation of several substances), and a furnace (which involves a chemical reaction).

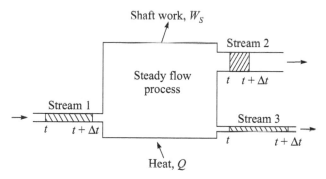

Figure 4.7 Energy balance on a steady flow process, like a turbine or a furnace or a liquid nitrogen plant. During the time interval Δt, each stream flows a certain distance, as represented schematically, and the fluid does work on the adjacent fluid.

W_s, called *shaft work*, includes any work exchanged between the equipment and the surroundings, with the exception of work done by the flowing fluid at the inlets and outlets. For a pump, the shaft work will be negative. For a turbine in an electric power-generating plant, W_s will include the electric work if the equipment is deemed to include the generator.

We should like the thermodynamic system to include the equipment and its instantaneous contents. This, then, is a simple example of an open system, one which is continually exchanging matter with its surroundings. For the analysis, however, we use a closed system including a given amount of material. At time t, the system includes the equipment, its contents, and some of the material in the inlet and outlet pipes, as indicated in Fig. 4.7. As the material flows, the boundary of this system moves. After an interval Δt, the crosshatched volume at the right side of Fig. 4.7 would have been added, while that on the left would have been deleted. Equation 4.9 applies to this system during the interval Δt.

For this process, Q has a simple meaning; it is the heat added to the system, as depicted in Fig. 4.7. The work W is the sum of shaft work W_s and the work done by the flowing fluid in the entrance and exit pipes. The fluid in stream 2, for example, must push out the fluid ahead of it before it can occupy the crosshatched volume. This involves the work $F_2 p_2 \Delta t / \rho_2$, where F_2 is the mass flow rate, p_2 is the pressure, and $\rho_2 = 1/\hat{V}_2$ is the density of stream 2. The work W thus will be

$$W = W_s + \Delta t \sum_s \frac{F_s p_s}{\rho_s}, \tag{4.30}$$

where the sum is over the several streams and F_s is positive for an exit stream and negative for an inlet stream.

The terms on the left in Eq. 4.9 represent the changes in the internal energy, potential energy, and kinetic energy of the thermodynamic system during the time interval Δt. Since this is a steady process, these changes are due solely to the replacement of the crosshatched volume on the left in Fig. 4.7 by those on the right. The total energy of the material in the equipment itself is the same at $t + \Delta t$ as at the time t. Equation 4.9 thus becomes

$$\Delta t \sum_s F_s \left(\hat{U}_s + gh_s + \frac{1}{2} v_s^2 \right) = Q - W = Q - W_s - \Delta t \sum_s \frac{F_s p_s}{\rho_s}, \tag{4.31}$$

where $\hat{U}_s$, gh_s, and $v_s^2/2$ represent the internal, potential, and kinetic energies, respectively, of stream s per unit mass. The definition of the enthalpy (per unit mass)

$$\widehat{H} = \widehat{U} + p\widehat{V} = \widehat{U} + p/\rho \qquad (4.32)$$

allows this result to be rewritten in the form

$$\sum_s F_s \left(\widehat{H}_s + gh_s + \frac{1}{2}v_s^2 \right) = \frac{Q - W_s}{\Delta t}, \qquad (4.33)$$

the right side representing the rate of heat addition minus the rate of shaft work.

Equation 4.33 is the result of the principle of conservation of energy as applied to an open system. Since the process is steady, this principle says that the net energy input must be zero. Contributions to this net energy input come from the heat and the shaft work as well as the energy carried by the flowing streams, and this is the meaning of Eq. 4.33. The only subtle feature is that there is a work term that combines with the internal energy in such a way that it appears that $\widehat{H}$ is the energy per unit mass carried by a moving fluid (in addition to the potential and kinetic energy). The effect here is similar to that encountered for constant-pressure processes. The enthalpy is an auxiliary material property that allows emphasis to be focused on thermal energy and forms of work other than pressure-volume work.

So many steady flow processes occur in engineering applications that Eq. 4.33 is used over and over again. It is worthwhile to write down two special cases. With emphasis on thermal energy, we frequently neglect the potential and kinetic energies. Equation 4.33 is then written in the symbolic but simple form

$$\Delta H = Q - W_s, \qquad (4.34)$$

it being borne in mind that ΔH is the difference in enthalpy between the streams flowing out of the process and those flowing in. A process with only one inlet and one outlet is the second special case. It is then unambiguous to write Eq. 4.33 as

$$\Delta \left(\widehat{H} + gh + \frac{1}{2}v^2 \right) = \widehat{Q} - \widehat{W}_s, \qquad (4.35)$$

since the mass flow rates of these two streams must be the same. $\widehat{Q}$ and $\widehat{W}_s$ are the heat and shaft work per unit mass of the flowing material.

Example 4.1 Water at 20°C, having a heat capacity of $\hat{C}_p = 1$ cal/g-K = 4.18 J/g-K, falls from a height of 30 m into a shallow basin. What will be the velocity of the water just above the basin, and what will be the temperature rise when the potential and kinetic energies have been converted into internal energy? Neglect the difference between $\hat{C}_p$ and $\hat{C}_V$, neglect evaporation, and assume that no work or heat is transmitted between the basin and the water. Neglect the drag force exerted by the atmosphere.

Solution: Let the water 30 m above the basin be state 1, water just above the basin be state 2, and quiescent water in the basin be state 3. The internal energy is the same in states 1 and 2. With $g = 9.8$ m/s², Eq 4.9 then yields

$$mg(h_2 - h_1) + \frac{1}{2}mv_2^2 = 0$$

or

$$v_2 = \sqrt{2g(h_1 - h_2)} = \sqrt{2 \times 9.8 \times 30} = 24.2 \text{ m/s}.$$

The velocity in state 3 is zero. Equation 9 again yields

$$U_3 - U_1 = mg(h_1 - h_3).$$

With no difference between $\hat{C}_p$ and $\hat{C}_V$, this becomes

$$m\hat{C}_p (T_3 - T_1) = mg(h_1 - h_3)$$

or

$$T_3 - T_1 = \frac{g(h_1 - h_3)}{\hat{C}_p} = \frac{9.8 \times 30}{4.18 \times 10^3} = 0.070 \,^\circ\text{C}.$$

For chemical processes, 30 m is a substantial height, and 24 m/s is a substantial velocity. Yet these correspond to only a very small temperature change. Consequently, kinetic and potential energies can frequently be ignored in comparison with thermal energy changes.

To be more rigorous, the velocity in state 2 should be calculated from momentum and force considerations. Equation 4.9 would then yield the change in kinetic energy.

Problems

4.1 Describe how the potential energy could be eliminated as a kind of energy by taking into account always a gravitational work term. There is a gravitational force exerted on a mass by other bodies, principally the earth. Is this the same as the force in Eq. 4.2?

4.2 Electric work was described in Eq. 4.10 as the product of an electric current, an electric potential, and an elapsed time. How can this be regarded as the product of a force and a distance?

4.3 Compare and contrast the following proposed statements of the first law of thermodynamics:

   (a) Energy is conserved.

   (b) Energy never disappears, although it may change from one form to another.

   (c) The internal energy U is a state function, that is, it is a property of the system.

4.4 (a) Does the first law of thermodynamics apply to irreversible processes?

   (b) Can Eq. 4.22 be applied to an irreversible process where the pressure is not uniform throughout the system, but only at its movable boundaries?

4.5 From the first law of thermodynamics, show for a PVT system that

$$dQ_{rev} = \left(\frac{\partial U}{\partial T}\right)_V dT + \left[\left(\frac{\partial U}{\partial V}\right)_T + p\right] dV$$

and show that dQ_{rev} is not an exact differential.

Notation

C_p heat capacity at constant pressure, J/K
C_V heat capacity at constant volume, J/K
E_k kinetic energy, J
E_p potential energy, J
E_{tot} total energy, J
F force, N

F_s mass flow rate, kg/s

g gravitational acceleration, m/s^2

h vertical position, m

H enthalpy, J

I electric current, A

l distance, m

m mass, kg

p pressure, N/m^2

Q heat added to system, J

t time, s

T absolute temperature, K

U internal energy, J

v velocity, m/s

V volume, m^3

W work done by the system, J

W' work, excluding pressure–volume work, J

W_s shaft work, J

ρ density, kg/m^3

Φ applied electric potential, V

Subscripts and special symbol

s stream

sat   saturated

^   (circumflex) per unit mass

Chapter 5

Thermal Properties of Gases

In this chapter, we intend to deal first of all with the heat capacity of gases in the limit of zero pressure, in the *ideal-gas state*. An asterisk is used to denote this state; thus, $\tilde{C}_p^*$ is the limit of the molar heat capacity $\tilde{C}_p$ at zero pressure. This is followed by a discussion of energy transformations involved in the compression, expansion, heating, and cooling of an ideal gas.

5.1 Heat Capacity

Chapter 4 included the definition of the heat capacity C_V at constant volume and the heat capacity C_p at constant pressure. These are related to the change in the internal energy of the material as the material exchanges heat and work with its surroundings. Energy is stored in the molecules of the material. Some of the internal energy can be attributed to kinetic energy of the molecules and some to potential energy. This is similar to macroscopic kinetic and potential energies, but it is not subject to the same kind of direct observation on the molecular level.

The first contribution to the heat capacity is related simply to the motion of the molecules as individual entities. The Maxwell–Boltzmann distribution applies to the velocities of the molecules of a substance. This was used in Problem 3.3 to show that the pressure of a gas might be attributed in substantial part to the motion of the

The Newman Lectures on Thermodynamics
John Newman and Vincent Battaglia
Copyright © 2019 Jenny Stanford Publishing Pte. Ltd.
ISBN 978-981-4774-26-0 (Hardcover), 978-1-315-10861-2 (eBook)
www.jennystanford.com

molecules. Similarly, the kinetic energy of the molecules can be obtained by integrating the velocity distribution:

$$
\left\{\begin{array}{l} \text{average} \\ \text{kinetic energy} \\ \text{per molecule} \end{array}\right\} = \frac{\iiint \frac{1}{2} m_i v^2 N(\underline{v}) dv_x dv_y dv_z}{\iiint N(\underline{v}) dv_x dv_y dv_z}, \tag{5.1}
$$

where the denominator assures that we have divided by the total number of molecules. The integration is over the three components of velocity and covers the entire range of possible velocities.

Since the integrands in Eq. 5.1 depend only on the magnitude of the velocity, it is convenient to transform $dv_x dv_y dv_z$ into a volume element in spherical coordinates, $v \sin\theta \, d\phi \, v \, d\theta \, dv$, where v is now integrated from zero to infinity, while θ and ϕ range from 0 to π and 0 to 2π, respectively. Integration over θ and ϕ thus yields a factor of 4π, which, however, cancels between the numerator and the denominator. Equation 5.1 now becomes

$$
\left\{\begin{array}{l} \text{average} \\ \text{kinetic energy} \\ \text{per molecule} \end{array}\right\} = \frac{\frac{1}{2} m_i \int_0^\infty v^4 e^{-m_i v^2/2kT} dv}{\int_0^\infty v^2 e^{-m_i v^2/2kT} dv} = kT \frac{\int_0^\infty x^4 e^{-x^2} dx}{\int_0^\infty x^2 e^{-x^2} dx}, \tag{5.2}
$$

where we have replaced $m_i v^2/2kT$ by x^2. Integration of the numerator by parts and multiplication by Avogadro's number L gives the result:

$$
\text{Molar translational internal energy} = \frac{3}{2} RT. \tag{5.3}
$$

The corresponding contribution to the molar heat capacity $\tilde{C}_V^*$ is $3R/2$ (see Eq. 4.11). This is sometimes regarded as the result of three *degrees of freedom* for translational motion, each degree contributing $R/2$ to the molar heat capacity.

For monatomic elements such as argon, neon, and krypton, we can stop with the translational contribution to the heat capacity. At sufficiently high temperatures, there would be an electronic contribution, due to the possibility of electrons occupying various energy states as more thermal energy becomes available. However, we shall not speak further of excited electronic states at the temperatures of interest in chemical processing.

Diatomic or linear molecules can also possess an appreciable kinetic energy by rotation in two independent directions. Rotation about the axis through the nuclei of the molecule does not make an appreciable contribution, however. For nonlinear, polyatomic molecules, rotation in three directions is important. Hence, linear molecules are said to possess two rotational degrees of freedom, and nonlinear molecules, three degrees. The corresponding contributions to the molar heat capacity are R for linear molecules and $3R/2$ for nonlinear molecules.

Carbon dioxide is a linear molecule, although it contains three atoms. The rotational contribution to the molar heat capacity for CO_2 thus is R, not $3R/2$.

Rotational energy levels are actually governed by quantum mechanics. Consequently, the rotational contribution to the heat capacity disappears at very low temperatures. In the temperature range of interest here, this effect can be ignored, and the contributions as stated above will apply.

Most interesting, perhaps, is the vibrational contribution to the heat capacity. A diatomic molecule is said to have one vibrational degree of freedom. In addition to their translational and rotational motion, the atoms can vibrate, alternately increasing and decreasing the interatomic distance. The vibrational energy alternates between kinetic energy of relative motion and potential energy of interatomic attraction. Because both kinetic and potential energies are involved, the one depending on the square of the velocity and the other on the square of the departure from a neutral position, the vibrational contribution to the molar heat capacity amounts to R, not $R/2$, for a fully excited vibrational degree of freedom.

Vibrations at the molecular level are also governed by quantum mechanics, and the vibrational contribution to the heat capacity goes to zero at low temperatures. However, vibrational levels are only *becoming* excited at ordinary temperatures, and this is the principal cause of the *variation* of the heat capacity with temperature. By contrast, the rotational energy is fully excited and the electronic energy is not at all active, for practical purposes.

The strength of a vibration is characterized by the force–distance relationship near the neutral position, and to a first approximation, other aspects of the interatomic force law can be ignored. This one constant determines the natural vibration frequency v, the spacing of the energy levels, and consequently the temperature at which thermal energy can be expected to become comparable to the spacing of the levels.

One consequence of this is that the heat capacities of several diatomic gases show a similar dependence on the temperature. (We restrict ourselves, for the moment, to diatomic gases so that there is only one vibrational mode to consider.) This similarity is made much more striking if we compare the heat capacities at the same value of a reduced temperature T/T_v, where T_v is a characteristic temperature for each diatomic gas and is related to the natural vibration frequency by

$$T_v = hv/k, \tag{5.4}$$

where h is Planck's constant (6.6256×10^{-35} J-s). Such a comparison is made in Fig. 5.1. Toward the left, the molar heat capacity $\tilde{C}_V^*$ approaches $5R/2$, corresponding to the translational and rotational contributions. Toward the right, $\tilde{C}_V^*$ approaches $7R/2$, the increase being caused by the activation of the vibrational degree of freedom.

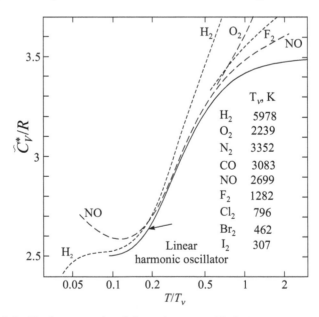

	T_v, K
H_2	5978
O_2	2239
N_2	3352
CO	3083
NO	2699
F_2	1282
Cl_2	796
Br_2	462
I_2	307

Figure 5.1 The heat capacity of diatomic gases, with the temperature reduced by the temperature $T_v = hv/k$, characteristic of the force of vibration between the atoms of the molecule. (Data courtesy: D. R. Stull and H. Prophet [2], who also give extensive data for polyatomic molecules.)

Observe that Fig. 5.1 constitutes a *corresponding states* correlation of the heat capacities $\tilde{C}_V^*$ of diatomic gases. As in the case of the compressibility factor in Section 3.6, the basis can again be found in molecular considerations. However, in the case of the PVT properties, the intermolecular force law was important (see Fig. 3.4), but for the heat capacity $\tilde{C}_V^*$, it was the interatomic force law within the molecule.

The frequency v can be measured directly by observation of the vibrational spectra of molecules. And with the aid of quantum mechanics, it is possible to predict the occupancy of various vibrational levels as a function of temperature. This means that it is possible to calculate Fig. 5.1 from fundamental principles and thereby avoid some delicate calorimetry on dilute gases. Although both the spectra and the calculations are more involved, the same method applies to polyatomic molecules. Molecular spectra are thus the source of the exact data for the heat capacities of gases in the ideal-gas state [1] and are so used for engineering purposes.

Translation, rotation, and vibration of molecules thus account for the heat capacity $\tilde{C}_V^*$ of gases. As shown in the second half of the chapter, the heat capacity at constant pressure is obtained by adding again the value of the universal gas constant

$$\tilde{C}_p^* = \tilde{C}_V^* + R, \tag{5.5}$$

so that $\tilde{C}_p^*$ thus has a similar dependence on temperature.

Our remarks have been restricted to the ideal-gas state, that is, the limit of the molar heat capacity at zero pressure. In this state, the gas obeys Eq. 3.8, and the molecules are essentially independent of each other in their translational kinetic energy and in the rotational and vibrational energies within the molecule. As the pressure is increased, the heat capacity will depart from the value $\tilde{C}_V^*$ or $\tilde{C}_p^*$ in the ideal-gas state. As the molecules are forced to become closer, there can be an additional potential energy of interaction between the molecules. Furthermore, the rotational and vibrational freedom of the molecules can be affected. This does not need to be of immediate concern to us because the pressure or volume dependence of the heat capacities can be obtained from the PVT properties. This requires the use of the second law of thermodynamics, the thermodynamic temperature scale, and the concept of entropy. Consequently, that discussion must be postponed until Chapter 7 and will be returned to again in Chapter 9.

The ideal-gas heat capacities (usually obtained from spectroscopic data) thus constitute fundamental data that can be used, with the PVT properties of Chapter 3, to construct complete thermodynamic information on a gas. For a number of gases, the temperature dependence of the molar heat capacity has been fitted to the equation

$$\tilde{C}_p^* = a + bT + c/T^2. \tag{5.6}$$

Values of the constants are given in Table 5.1.

Table 5.1 Constants in Eq. 5.6 for the heat capacity of gases at zero pressure.

Gas	a (cal/K-mol)	$10^3 b$ (cal/K²-mol)	$10^{-5}c$ (cal-K/mol)
Monatomic	4.97	0	0
S	5.26	−0.10	0.36
H_2	6.52	0.78	0.12
O_2	7.16	1.00	−0.40
N_2	6.83	0.90	−0.12
S_2	8.72	0.16	−0.90
CO	6.79	0.98	−0.11
F_2	8.26	0.60	−0.84
Cl_2	8.85	0.16	−0.68
Br_2	8.92	0.12	−0.30
I_2	8.94	0.14	−0.17
CO_2	10.57	2.10	−2.06
H_2O	7.30	2.46	0
H_2S	7.81	2.96	−0.46
NH_3	7.11	6.00	−0.37
CH_4	5.65	11.44	−0.46

Source: Ref. [1], p. 66.
Note: The fit is valid from 298 to 2000 K.

5.2 Processes Involving Ideal Gases

An ideal gas obeys Eq. 3.8

$$p\tilde{V} = RT, \tag{5.7}$$

and its heat capacity

$$\tilde{C}_V^* = \left(\frac{\partial \tilde{U}}{\partial T}\right)_V \tag{5.8}$$

has been discussed in the first part of this chapter. We shall further assume that the internal energy depends only on the temperature, that is

$$\left(\frac{\partial U}{\partial V}\right)_T = 0. \tag{5.9}$$

This implies that $\tilde{C}_V^*$ depends on T only, in harmony with our definition of $\tilde{C}_V^*$ as the zero-pressure limit of $\tilde{C}_V$.

It is a good assumption for the internal energy to be independent of volume as long as the molecules are far apart and do not interact strongly with each other. This is, in fact, the same condition under which the ideal-gas law applies. Equation 5.9 is obeyed by real gases in the limit of zero pressure. It will be possible to show in Chapter 7 or 9 (see Eqs. 7.13 and 9.4) that Eq. 5.9 follows exactly for a substance that obeys Eq. 5.7 exactly, and deviations from both relations are related to each other.

In the meantime, the introduction of Eqs. 5.7–5.9 allows us to treat some interesting processes without having to wait for the development of the complete thermodynamic framework for real substances. No real substance is an ideal gas. For some processes, the ideal-gas approximation gives useful results. In other situations, qualitatively different results are obtained for real substances. For example, $(\partial \tilde{U}/\partial p)_T$ does not approach zero in the limit of zero pressure even though zero is the correct limit for $(\partial \tilde{U}/\partial \tilde{V})_T$. In fact, $(\partial \tilde{U}/\partial p)_T$ is typically much larger for real gases than for real liquids. Attention will be called to some of these examples in subsequent chapters. One of the truncated virial Eqs. 3.24 or 3.25 is useful for assessing departures from ideal-gas behavior without introducing unnecessary complication.

For an ideal gas, the molar enthalpy is

$$\tilde{H} = \tilde{U} + p\tilde{V} = \tilde{U} + RT. \tag{5.10}$$

Since $\tilde{U}$ is a function of T only, it follows that the enthalpy $\tilde{H}$ also depends only on temperature. In particular, the heat capacity $\tilde{C}_p^*$ at constant pressure is (see Eq. 4.24)

$$\tilde{C}_p^* = \left(\frac{\partial \tilde{H}}{\partial T}\right)_p = \left(\frac{\partial \tilde{U}}{\partial T}\right)_p + R. \tag{5.11}$$

In view of Eq. 5.9,

$$\left(\frac{\partial \tilde{U}}{\partial T}\right)_p = \left(\frac{\partial \tilde{U}}{\partial T}\right)_V + \left(\frac{\partial \tilde{U}}{\partial \tilde{V}}\right)_T \left(\frac{\partial \tilde{V}}{\partial T}\right)_P = \left(\frac{\partial \tilde{U}}{\partial T}\right)_V. \tag{5.12}$$

Hence, Eq. 5.11 reduces to Eq. 5.5 for the relationship between the molar heat capacities at constant pressure and at constant volume, for an ideal gas.

For reversible processes, we can take the work to be given by $đW = pdV$ (see Eq. 3.1). The first law of thermodynamics (Eq. 4.15) then takes the form

$$dU = đQ - pdV. \tag{5.13}$$

Since U depends only on temperature for an ideal gas, we finally have a way to calculate the heat directly for any reversible process involving an ideal gas:

$$đQ = dU + pdV = C_V^* \, dT + pdV. \tag{5.14}$$

When pressure is a more convenient independent variable, this result can be rewritten with Eqs. 5.7 and 5.5 to yield

$$đQ = C_V^* dT + nRdT - \frac{nRT}{p}dp = C_p^* dT - \frac{nRT}{p}dp. \tag{5.15}$$

A process carried out at constant temperature would appear to convert heat directly into work since then $đQ = đW$ for an ideal gas.

For a constant-pressure process,

$$Q = \int_{T_1}^{T_2} C_p^* dT = \Delta H \quad \text{and} \quad W = p\Delta V. \tag{5.16}$$

From this, it follows that

$$\Delta T = \frac{p\Delta\tilde{V}}{R} \quad \text{and} \quad \Delta U = \int_{T_1}^{T_2} C_V^* dT . \tag{5.17}$$

At constant volume,

$$W = 0 \quad \text{and} \quad Q = \int_{T_1}^{T_2} C_V^* dT = \Delta U. \tag{5.18}$$

Consequently,

$$\Delta T = \frac{\tilde{V}\Delta p}{R} \quad \text{and} \quad \Delta H = \int_{T_1}^{T_2} C_p^* dT . \tag{5.19}$$

A reversible, adiabatic process, where $Q = 0$, can involve a simultaneous variation of T, p, and V. It is customary to define γ as the ratio of the specific heats at constant pressure and at constant volume:

$$\gamma = \frac{C_p^*}{C_V^*} = 1 + \frac{R}{\tilde{C}_V^*} . \tag{5.20}$$

Equation 5.14 can also be written in the form

$$dQ = \frac{\tilde{C}_V^*}{R} (Vdp + \gamma p dV). \tag{5.21}$$

When γ can be approximated as a constant, a reversible, adiabatic process, therefore, follows the relationship

$$pV^\gamma = \text{constant}. \tag{5.22}$$

Equivalent forms are

$$TV^{\gamma-1} = \text{constant} \tag{5.23a}$$

and

$$T^\gamma p^{1-\gamma} = \text{constant}. \tag{5.23b}$$

One or another of these equations would be appropriate for compression by a given volume ratio, a given pressure ratio, or a given temperature ratio. The work can be computed when the temperature change is known

$$W = Q - \Delta U = -\Delta U = C_V^* (T_1 - T_2). \tag{5.24}$$

Adiabatic compressions and expansions are more complex to analyze if the temperature dependence of the heat capacity, or of γ, must be recognized. If the final temperature is not given, a trial-and-error calculation is involved. Integration of Eq. 5.15 for an adiabatic process gives

$$\int_{T_1}^{T_2} \frac{C_p^*}{R} dT = nR \ln \frac{p_2}{p_1} .$$ (5.25)

For a given pressure ratio, it is more difficult to compute T_2 from this relation than from Eq. 5.23b. Working on Problem 5.1 given at the end of the chapter gives an opportunity to employ one or more degrees of exactness in the compression of a gas to an unknown final temperature.

Some irreversible processes can be treated when the nature of the irreversibility is adequately described. Example 5.3 treats the mixing of two portions of gas initially at different temperatures and pressures. In this particular situation, both the work and the heat are zero.

Steady-flow processes deserve special mention here because of their importance in engineering systems. Recall that conservation of energy was expressed especially for such processes in the last part of Chapter 4. For reversible-flow processes, the work is still given by Eq. 3.1, if one can follow individual fluid elements through the equipment. For a system with one inlet and one outlet, the work associated with moving the fluid at the inlet and the outlet, per unit mass, can be written as $\Delta p \widehat{V}$ (compare Eq. 4.30). Consequently, the shaft work is

$$\widehat{W}_s = \int p dV - \Delta p \widehat{V} = -\int \widehat{V} dp .$$ (5.26)

This equation is not restricted to ideal gases. On a p–V diagram, such as Fig. 5.2 or Fig. 3.1, the integral in Eq. 5.26 represents the area to the left of the process curve, while the work $\widehat{W}$ is the area below the process curve. These two areas are equal for the isothermal compression or expansion of an ideal gas, but otherwise they are generally different.

Flow processes are similar to nonflow processes, except that W_s is now more relevant than W. For example, for a reversible, adiabatic process where kinetic and potential energy changes can be ignored,

Eqs. 5.22 and 5.23 can be applied. But instead of Eq. 5.24, we would write

$$\widehat{W}_s = \widehat{Q} - \Delta\widehat{H} = -\Delta\widehat{H} = \widehat{C}_p^*\left(T_1 - T_2\right). \qquad (5.27)$$

Since $\widehat{C}_p^*$ is greater than $\widehat{C}_V^*$, the shaft work for continuous compression of a gas is greater than one would compute as the work W for a nonflow process.

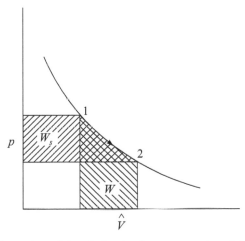

Figure 5.2 Illustration of the distinction between work and shaft work for a reversible flow process.

Examples 5.1 and 5.2 treat a flow process, with comparison of a reversible process and a real process.

Example 5.1 An adiabatic compressor in a steam loop has no instruments that indicate directly the flow rate of the material being compressed. However, instruments do indicate the temperature and pressure of the material at both the inlet and the outlet, as well as the electric power being consumed by the compressor. Treat the steam as an ideal gas with a heat capacity $\widehat{C}_p$ = 0.465 cal/g-°C and a ratio of heat capacities $\gamma = \widehat{C}_p/\widehat{C}_V$ = 1.31 and estimate the flow rate in the case where the power consumption is 1 kW and the readings are

	Inlet	Outlet
Pressure (psia)	35	75
Temperature (°F)	400	600

Solution: For a steady, adiabatic process, the shaft work is related to the enthalpy change for the flowing fluid:

$$W_s = -\Delta H.$$

For an ideal gas, the enthalpy depends only on the temperature:

$$\Delta \widehat{H} = \widehat{C}_p \, (T_2 - T_1) = 0.465 \times 200/1.8 = 51\frac{2}{3} \text{ cal/g}.$$

Hence, the flow rate is given by

$$\frac{1000 \text{ J}}{\text{s}} \frac{\text{cal}}{4.184 \text{ J}} \frac{1.8}{200 \times 0.465} \frac{\text{g}}{\text{cal}} = 4.626 \text{ g/s}.$$

Example 5.2 In the situation of Example 5.1, calculate the ratio of the actual work to the work that would have been required for a reversible adiabatic compression to the same final pressure.

Solution: Since $W_s = -\Delta H$ for both the actual and the reversible processes and since, for an ideal gas, the enthalpy depends only on the temperature, the required ratio can be expressed as

$$\frac{W_s}{W_{s,\text{rev}}} = \frac{\Delta H}{\Delta H_{\text{rev}}} = \frac{\widehat{C}_p(T_2 - T_1)}{\widehat{C}_p(T_2 - T_1)_{\text{rev}}} = \frac{T_2 - T_1}{(T_2 - T_1)_{\text{rev}}}.$$

Thus, we need only $T_{2,\text{rev}}$, the final temperature for the reversible adiabatic compression to the same final pressure. For such a process involving an ideal gas with a constant ratio γ of heat capacities, we can write

$$p\widetilde{V}^{\gamma} = \text{const} = p(RT/p)^{\gamma}$$

or

$$T_{2,\text{rev}} = T_1(p_2/p_1)^{\frac{\gamma-1}{\gamma}} = T_1(75/35)^{0.31/1.31} = 1.19764\,T_1.$$

Since $T_1 = 400 + 459.67 = 859.67°R$, we obtain $T_{2,\text{rev}} = 1029.576°R = 569.91°F$, and substitution into the equation for the ratio of the work yields

$$\frac{W_s}{W_{s,\text{rev}}} = \frac{200}{169.91} = 1.177.$$

Example 5.3 Two masses of nitrogen, which can be treated as an ideal gas, are mixed irreversibly under the conditions depicted in Fig. 5.3 by the removal of the separating diaphragm. The heat capacity of nitrogen can be taken to be constant. Calculate the final temperature and pressure of the nitrogen.

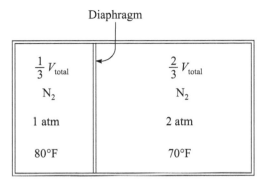

Diaphragm

$\frac{1}{3} V_{total}$

N_2

1 atm

80°F

$\frac{2}{3} V_{total}$

N_2

2 atm

70°F

Figure 5.3 Irreversible mixing of an ideal gas. The outer walls are rigid and insulating.

Solution: Let the initial left side be denoted by 1 and the initial right side by 2. A mass balance, an energy balance, and a PVT calculation are essential in the determination of the final temperature and pressure. First let us calculate the initial number of moles on each side.

$$n = n_1 + n_2$$

$$n_1 = \frac{V/3}{539.67R} = 0.000618 \frac{V}{R}, \text{ since } p_1 = 1 \text{ atm}$$

$$n_2 = \frac{2 \times 2V/3}{529.67R} = 0.002517 \frac{V}{R}, \text{ since } p_2 = 2 \text{ atm}$$

Hence $V/Rn = 318.984°R/atm$, and $n_1/n = 0.197$ and $n_2/n = 0.803$.

The final temperature is now determined by the condition that the internal energy does not change for this process, $\Delta U = 0$ or

$$(n_1 + n_2)\tilde{C}_V T_f = n_1 \tilde{C}_V T_1 + n_2 \tilde{C}_V T_2$$

or

$$T_f = 0.197 \times 80 + 0.803 \times 70 = 71.97°F = 531.64°R$$

since the internal energy depends only on the temperature for an ideal gas.

The final pressure is now given by

$$p_f = \frac{nRT_f}{V} = \frac{531.64}{318.984} = 1.6667 = \frac{5}{3}\,\text{atm}.$$

Problems

5.1 In the process of degassing an oil stream of methane, a suitable vacuum pump draws in methane at 0.05 atm and 25°C, compresses it reversibly and adiabatically, and discharges it at 1 atm pressure. Determine the work done per mole of methane compressed and discharged, as well as the temperature of the methane leaving the compressor. Under the conditions of the process, methane may be assumed to behave as an ideal gas having a molar heat capacity given by $\tilde{C}_p^*/R$ = 2.46 + 0.006T, where T is expressed in Kelvin. As a challenge, try to avoid assuming that the heat capacity is constant.

5.2 On the basis of Eq. 5.26, explain why it should be more expensive to move a gas than to pump a liquid against a similar back pressure.

5.3 In the adiabatic compression of an ideal gas from 1 atm to 5 atm, how much work is required? The actual work is 1.25 times larger than the reversible work, and the heat capacity $\tilde{C}_p^*$ is 5 Btu/lb-mole-°R.

Notation

a, b, c	constants in Eq. 5.6
C_p	heat capacity at constant pressure, J/K
C_V	heat capacity at constant volume, J/K
h	Planck's constant, 6.6256 × 10^{-34} J-s
H	enthalpy, J
k	Boltzmann's constant, 1.38 × 10^{-23} J/K
L	Avogadro's (or Loschmidt's) number, 6.0225 × 10^{23} mol^{-1}
m_1	molecular mass, kg
n	number of moles
N	distribution function, s^3/m^3
p	pressure, N/m^2
Q	heat, J

R	universal gas constant, 8.3143 J/mol-K or 82.06 atm-cm^3/mol-K
T	absolute temperature, K
$T_v = h v/k$	temperature characteristic of molecular vibration, K
U	internal energy, J
υ	velocity, m/s
V	volume, m^3
W	work, J
W_s	shaft work, J
x, y, z	rectangular coordinates, m
$\gamma = C_p^*/C_V^*$	ratio of heat capacities
θ	angular coordinate
v	natural vibration frequency, s^{-1}
ϕ	angular coordinate

Superscript and special symbols

*	ideal-gas state
~	per mole
^	per unit mass

References

1. Gilbert Newton Lewis and Merle Randall, revised by Kenneth S. Pitzer and Leo Brewer, *Thermodynamics* (New York: McGraw-Hill Book Company, Inc., 1961), Chapter 27.

2. D. R. Stull and H. Prophet, *JANAF Thermochemical Tables*, Second Edition, June, 1971, NSRDS-NBS 37.

Chapter 6

Second Law of Thermodynamics

The equivalence of heat and work according to the first law of thermodynamics is now almost universally accepted. Separate units for heat, such as caloric and the British thermal unit, are falling into disuse, and units for work (or energy) are used instead. Once the first law is mastered, the primary concern of thermodynamics becomes the second law and its consequences. How well do we understand why it is necessary to use about 10 MJ of thermal energy to produce 1 kWh or 3.6 MJ of electrical energy?

In this chapter, certain generalizations of experimental observations are translated into the second law of thermodynamics. One of these starting points is the observation that heat always flows spontaneously from a region of high temperature to a region of low temperature. The logic of the development is sound, but its comprehension requires some attention on the part of the readers. They must understand the Carnot cycle, how this cycle is used to define the thermodynamic temperature scale, and how these concepts lead to the entropy, a new property of the system.

There are other routes of logic that lead to the new system property, entropy, but they are even more abstract. The second law was originally developed by means of the Carnot cycle, basically as outlined here, and the development should be appealing to engineering students because of its involvement with heat engines

The Newman Lectures on Thermodynamics
John Newman and Vincent Battaglia
Copyright © 2019 Jenny Stanford Publishing Pte. Ltd.
ISBN 978-981-4774-26-0 (Hardcover), 978-1-315-10861-2 (eBook)
www.jennystanford.com

and because it illustrates the creative thought patterns actually used historically.

Our concern with the second law does not end with the concept of entropy. This merely puts us in a position to consider the consequences of the second law. These include limitations on the efficiency of conversion of thermal energy to mechanical energy. But they also range over diverse topics such as the formation of two coexisting phases instead of one, the minimum energy required to separate components of a solution, the equilibrium composition of a reactive gas mixture, the electric potential of a galvanic cell, and adsorption at an interface between two phases. The development of such applications will occupy the remainder of the book.

6.1 Irreversible Processes

We like to treat quasi-static processes in thermodynamics. Then the system is always close to thermodynamic equilibrium, and properties such as temperature and pressure can be regarded as uniform throughout the system.[1] A quasi-static process should be reversible, that is, it should be able to go equally well in either direction, and the "losses" should become more nearly negligible as the process becomes more nearly static.

The second law of thermodynamics has this to say about reversible and irreversible processes: in going through a process, the "losses" are always zero or positive.

Let us consider what we mean by losses, using a gas as an example system. Figure 6.1 shows the isotherms on a graph of pressure versus volume. The work of compressing a gas for a nearly isothermal process is obtained by integrating the expression

$$\dbar W = p\,dV = \frac{nRT}{V}dV \tag{6.1}$$

to yield

$$W = \int_1^2 \frac{nRT}{V}dV = nRT\ln\frac{V_2}{V_1}. \tag{6.2}$$

---

[1]The rate laws for irreversible processes are treated in fluid mechanics, chemical kinetics, heat transfer, and mass transfer (diffusion). Departures from thermodynamic equilibrium provide the driving force for the rate process, but we are not yet prepared to treat the resistance to the process occurring.

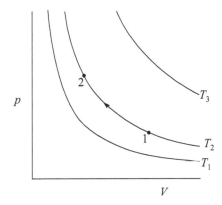

Figure 6.1 Pressure–volume plot for a gas, depicting compression at constant temperature.

W is negative, because work is done on the system. (Does the work depend on the manner of going from state 1 to state 2, even for a reversible process? Is energy conserved if the work is not the same but the final states are?)

Suppose the process is not carried out reversibly. Do your instinct and experience tell you that the pressure required will be greater or less than the reversible pressure? They should tell you that a greater pressure is required, and this is in harmony with the second law of thermodynamics. For example, slow heat transfer to the walls of the chamber will result in a somewhat higher average temperature, and this corresponds to a pressure higher than the reversible pressure. Similarly, shockwaves formed by rapid movement of the piston will result in a higher pressure on compression.

Now expand the gas back to the initial state. Instinct and experience tell us that p is now less than the reversible value and that the work done by the system is less than the reversible value. We now have a complete but irreversible cycle, which is completed at the point where we started (see Fig. 6.2). The net work done by the system in this cycle is negative (work was done on the system). What happened to this work? Is the first law of thermodynamics obeyed? The second law of thermodynamics tells us that the loss is always in this direction. We lose work but gain heat. Mechanical or electrical energy is irreversibly converted to thermal energy. We conclude that energy in the form of heat or internal energy is not as useful as mechanical energy.

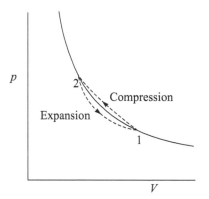

Figure 6.2   Irreversible cycle of compression and expansion.

6.2   Heat Engines and Carnot's Theorem

We want to consider various power cycles for converting internal energy into work. These are all like those sketched in Fig. 6.3.

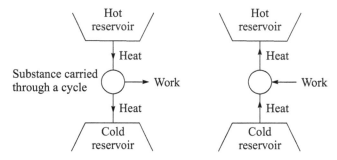

Figure 6.3   Heat engine (left) and refrigerator (right).

The *working substance* is carried through a cyclic process in order that its state may be the same at the end as at the beginning. The above process can produce work, or it can be run backwards to transfer heat from a cold reservoir to a hot reservoir (a refrigerator).

Applied to such processes, the first law of thermodynamics says

$$\Delta U = 0 = Q_H + Q_C - W. \qquad (6.3)$$

The cyclic nature of the process dictates that ΔU is zero. The sign conventions for heat and work are indicated in Fig. 6.4—heat added to the working substance is positive, and work done by the working substance is positive.

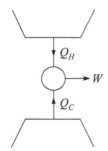

Figure 6.4 Sign conventions for heat and work.

The *efficiency* of a power cycle is defined as

$$\eta = \frac{W}{Q_H} = \frac{Q_H + Q_C}{Q_H} = 1 + \frac{Q_C}{Q_H} = 1 - |Q_C/Q_H|. \qquad (6.4)$$

This definition also applies to a refrigerator; however, the *coefficient of performance* (COP),

$$\text{COP} = -Q_C/W \qquad (6.5)$$

reflects more accurately what the refrigerator is supposed to accomplish.

Our experience governing the conversion of work to heat and *vice versa* is embodied in the second law of thermodynamics, expressed as follows:

1. It is impossible to construct an engine that, operating in a cycle, will produce no effect other than the extraction of heat from a reservoir and the performance of an equivalent amount of work (in other words, we need a flow $-Q_C$ to the cold reservoir and η must be less than unity)

   or

2. It is impossible to construct a device that, operating in a cycle, will produce no effect other than the transfer of heat from a cooler to a hotter body (in other words, the work $-W$ is needed in order to operate the refrigerator in Fig. 6.3).

These two statements are equivalent, and they apply to any of our power cycles and refrigerators.

Carnot came up with the simplest such cyclic process, completing the cycle with two isothermal processes and two adiabatic processes, all four processes being reversible. By this selection of processes, Carnot's cycle requires only one reservoir at the higher temperature T_H and one reservoir at the lower temperature T_C. During the nonisothermal adiabatic processes, the substance is isolated from any reservoir. A Carnot cycle is sketched in Fig. 6.5 for a PVT system in the gaseous state. However, it is important that the working substance need not be a PVT system. It could be a surface system, an electrochemical cell, a magnetic substance, etc.

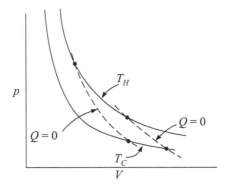

Figure 6.5 Carnot cycle for a PVT system in the gaseous state.

Carnot proved the theorem that "No engine operating between two given reservoirs can be more efficient than a Carnot engine operating between the same two reservoirs." This is proved by showing that the existence of such an engine would violate the second law of thermodynamics (see Fig. 6.6). We use the supposed engine, I, to operate a Carnot refrigerator; the two together constitute a device that uses no work and produces no work. The compound device merely transfers heat between the reservoirs.

The hypothesis $\eta_I > \eta_C$ implies that since W is same for the two systems,

$$\frac{W}{|Q_{H'}|} > \frac{-W}{-|Q_H|} \tag{6.6}$$

or

$$|Q_H| > |Q_{H'}|. \tag{6.7}$$

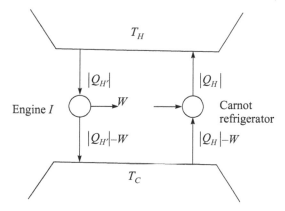

Figure 6.6 Sketch used to prove Carnot's theorem.

The combined process thus transfers a net amount of heat from the cold reservoir to the hot reservoir, in violation of the second law of thermodynamics.

The reversibility of the Carnot cycle is crucial in this proof since otherwise the efficiency of a Carnot refrigerator would be different from the efficiency of the Carnot engine operating between the same two reservoirs.

6.3   Thermodynamic Temperature Scale

It follows from Carnot's theorem that all Carnot engines are equally efficient if operated between the same reservoirs. Thus, the efficiency of the Carnot engine is independent of the working substance and independent, in fact, of whether a PVT system is used to construct the Carnot engine. (The possibility of using a surface system or an electrochemical system for the Carnot cycle is considered in Chapter 10.)

This means that the efficiency of the Carnot engine must depend only on some characteristics of the *reservoirs*, and the only characteristic with which the reservoirs have been endowed is *temperature*. This provides us with an opportunity to define a temperature scale, one which is now independent of the working substance of the thermometer. (In Chapter 2, this was considered to be one desirable criterion for a temperature scale.) Recall that the ideal-gas temperature scale, in the limit of zero pressure, was independent of which gaseous substance was used. When a Carnot engine is the thermometer, the thermometric substance does not even need to be a gas.

Let us suppose that we have defined a temperature scale θ, following one of the methods discussed in Chapter 2. This scale should be reproducible, but it does not need to have any other special characteristic. Consider now three reservoirs and two Carnot engines, as shown in Fig. 6.7. Adjust the heat loads for the two engines so that Q_C is the same for the two engines and thus no net heat is transferred to the intermediate reservoir at temperature θ_C.

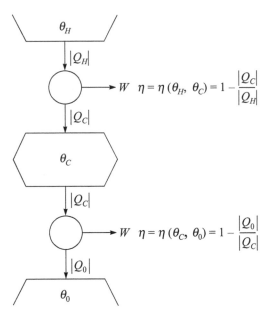

Figure 6.7 Two Carnot engines between reservoirs at temperatures θ_H, θ_C, and θ_0 can be regarded as one Carnot engine between the two reservoirs at temperatures θ_H and θ_0.

The two Carnot engines taken together constitute a single Carnot engine operating between the reservoirs at temperatures θ_H and θ_0. This Carnot engine has the efficiency

$$\eta = 1 - \frac{|Q_0|}{|Q_H|} = \eta(\theta_H, \theta_0). \qquad (6.8)$$

The efficiencies of the component engines are indicated in Fig. 6.7.

Two of these equations can be rearranged to read

$$\frac{|Q_0|}{|Q_C|} = 1 - \eta(\theta_C, \theta_0) \tag{6.9}$$

and

$$\frac{|Q_0|}{|Q_H|} = 1 - \eta(\theta_H, \theta_0), \tag{6.10}$$

and the ratio of these two is directly related to the third:

$$\frac{|Q_C|}{|Q_H|} = \frac{1 - \eta(\theta_H, \theta_0)}{1 - \eta(\theta_C, \theta_0)} = 1 - \eta(\theta_H, \theta_C). \tag{6.11}$$

This result tells us that the ratio in the middle of Eq. 6.11 must really be independent of θ_0 (since the right side does not depend on θ_0), and it can be written as a ratio of a function of θ_H to this same function of θ_C

$$-\frac{Q_C}{Q_H} = \frac{T(\theta_C)}{T(\theta_H)} = 1 - \eta(\theta_H, \theta_C), \tag{6.12}$$

where

$$T(\theta_C) = \frac{\text{constant}}{1 - \eta(\theta_C, \theta_0)} \tag{6.13}$$

and

$$T(\theta_H) = \frac{\text{constant}}{1 - \eta(\theta_H, \theta_0)}. \tag{6.14}$$

(This constant can depend on θ_0, but it cancels in Eq. 6.12.)

Equation 6.12 defines the thermodynamic temperature scale T. For any Carnot engine operating between reservoirs at the temperature θ_H and θ_C, the ratio of the thermodynamic temperatures $T_C = T(\theta_C)$ and $T_H = T(\theta_H)$ is given by Eq. 6.12, either in terms of the efficiency $\eta(\theta_H, \theta_C)$ or in terms of the ratio of heats Q_C/Q_H of the cycle. The definition is completed by specifying the temperature $T_t = 273.16$ K at the triple point of water, and the resulting scale is called the Kelvin temperature scale.

The equivalence of the thermodynamic temperature scale to the ideal-gas temperature scale is demonstrated in Chapter 9. (See also Problem 6.3.) However, the thermodynamic temperature scale is necessary for the development of thermodynamic relationships in Section 6.4 and in Chapter 7. It also finds practical application in the measurement of temperatures near absolute zero, beyond the range of the ideal-gas thermometer. In other words, any substance that can perform a Carnot cycle can be used to measure temperature. Magnetic properties of matter are used in this way for the very low temperatures.

From the construction of the thermodynamic temperature scale, one sees that the efficiency of a Carnot cycle operating between reservoirs at the temperatures T_H and T_C now takes the simple form

$$\eta = 1 - \frac{T_C}{T_H}. \qquad (6.15)$$

We could have been more simple minded in our approach to thermometry with Carnot engines. We could have picked a primary reference temperature, for example, the temperature θ_t at the triple point of water. Then, the efficiency $\eta(\theta, \theta_t)$ of a Carnot engine operated between a reservoir at θ and a reservoir at θ_t would constitute an unambiguous thermometric property. We can go beyond this by using the thought process depicted in Fig. 6.7 and the development in the text to eliminate the special place of the temperature θ_t. For example, suppose we have measured the thermodynamic temperature of the melting point of sulfur by means of a Carnot engine operating between this temperature and θ_t. From measurements on a Carnot engine operated between a different temperature θ and the sulfur point, can we determine the thermodynamic temperature $T(\theta)$ or must we always use a reservoir at θ_t? This is the question we have answered in Eq. 6.12. A similar question for thermocouples is treated in Example 6.1.

6.4 A New State Property: Entropy

For a Carnot cycle, that is, any thermodynamic system operated in a reversible cycle between only two reservoirs T_H and T_C (see Fig. 6.5), we have

$$\frac{Q_H}{Q_C} = -\frac{T_H}{T_C} \qquad (6.16)$$

or

$$\frac{Q_H}{T_H} + \frac{Q_C}{T_C} = 0 . \tag{6.17}$$

Therefore, for this reversible cycle,

$$\oint_{\text{rev}} \frac{dQ}{T} = 0 . \tag{6.18}$$

We can extend this theorem to any reversible cycle by approximating it by a series of Carnot cycles, as shown in Fig. 6.8. $Q = 0$ for all the adiabatic processes, and

$$\sum \frac{Q_H}{T_H} + \frac{Q_C}{T_C} = 0 \tag{6.19}$$

for all the Carnot cycles. Any reversible cycle can be represented exactly in the limit of an infinite number of Carnot cycles, and we have

$$\oint_{\text{rev}} \frac{dQ}{T} = 0 . \tag{6.20}$$

It follows that dQ_{rev}/T is the differential of a state function called *entropy* and is given by the symbol S:

$$\frac{dQ_{\text{rev}}}{T} = dS \tag{6.21}$$

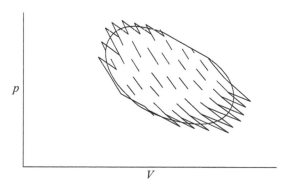

Figure 6.8 Any reversible cycle can be approximated by segments of isothermal processes and adiabatic processes, and these can be grouped into equivalent Carnot cycles.

or

$$S_2 - S_1 = \int_{state\,1}^{state\,2} \frac{dQ_{rev}}{T} = \int_1^2 dS \qquad (6.22)$$

or

$$\oint dS = 0. \qquad (6.23)$$

That is, the value of S_2-S_1 is independent of the path from state 1 to state 2, or the value of the integral over any closed path is zero. These are essential features of any state function.

Notice that dQ_{rev} is still *not* the differential of a state function. In the same way, dW_{rev}/p is the differential of the state function V for a PVT system, although dW_{rev} itself is still not the differential of a state function.

For reversible processes with a PVT system,

$$dW = pdV \qquad (6.24)$$

and

$$dQ = TdS. \qquad (6.25)$$

Hence, the first law of thermodynamics

$$dU = dQ - dW \qquad (6.26)$$

becomes

$$dU = TdS - pdV. \qquad (6.27)$$

We see that this equation now involves only state variables and can, therefore, be applied to any process, reversible or irreversible. For example, suppose that an irreversible process takes a PVT system from state 1 to state 2. The internal energy change for the process can be calculated by integrating Eq. 6.27 for any reversible process that also takes the system from state 1 to state 2. This latter process should be reversible in order to ensure that the properties such as T, p, and $\hat{S}$ are well defined and uniform throughout the system.

From Eq. 6.27, it follows that

$$T = \left(\frac{\partial U}{\partial S}\right)_V. \qquad (6.28)$$

This should emphasize that T is the thermodynamic temperature defined in Section 6.3, which only by coincidence is equal to the ideal-gas temperature. What else can we do with our new knowledge? The consequences of the entropy function in relating the properties of PVT systems are developed in Chapter 7.

This chapter has included several statements of the second law of thermodynamics:

1. The "losses" in a process are always positive or zero. (See Section 6.1.)
2. It is impossible to construct an engine that, operating in a cycle, will produce no effect other than the extraction of heat from a reservoir and the performance of an equivalent amount of work.
3. It is impossible to construct a device that, operating in a cycle, will produce no effect other than the transfer of heat from a cooler to a hotter body.
4. There exists "entropy," which is a (state) property of a thermodynamic system and whose differential is dQ_{rev}/T. This fact was derived from statement 3 in the present section. As an exercise, derive statement 2 from statement 4 (see Problem 8.3).
5. For any real or ideal process, the entropy change of the system and the surroundings is positive or zero, being zero only in ideal processes. This statement, which is similar to statement 1, is developed in Chapter 8.

Example 6.1 Copper–constantan thermocouples are used to measure temperatures in a distillation column, and a table is provided of the potential $\Phi(T, T_0)$ of such a thermocouple as a function of the temperature T of one junction when the other junction is maintained at the fixed temperature of $T_0 = 0°C$. On this particular day, however, no ice can be found to maintain the reference junction at 0°C. Instead, the reference junction is maintained at room temperature, which a mercury thermometer records to be 23°C. How should we use the measured thermocouple potentials and the table of $\Phi(T, T_0)$ to determine the temperature of the other junction?

Solution: Let the reference-junction temperature be denoted by T_C and the sought temperature be denoted by T_H (see Fig. 6.9). The measured potential thus is $\Phi(T_H, T_C)$. Now imagine that a second thermocouple is set up between the temperature $T_C = 23°C$ and $T_0 = 0°C$. We can look up the potential $\Phi(T_C, T_0)$ of this second, hypothetical thermocouple in our table. The two junctions at T_C can now be connected by a constantan wire, the dashed line in Fig. 6.9. The two thermocouples now constitute a single thermocouple between the temperatures T_H and T_0, and it is apparent that the potentials are related by

$$\Phi(T_H, T_0) = \Phi(T_H, T_C) + \Phi(T_C, T_0).$$

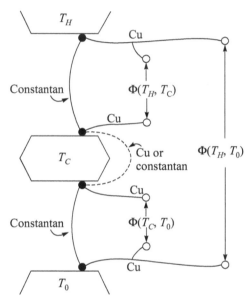

Figure 6.9 System of copper–constantan thermocouples and three reservoirs. Solid dots represent thermocouple junctions. The dashed line can be a wire of either copper or constantan.

The two values on the right are now known. We thus calculate $\Phi(T_H, T_0)$ and look up the temperature T_H in the table.

We have shown that

$$\Phi(T_H, T_C) = \Phi(T_H, T_0) - \Phi(T_C, T_0)$$

is the difference between a function of T_H and a function of T_C, the result being independent of T_0. Thus,

$$\Phi(T_H, T_C) = \psi(T_H) - \psi(T_C).$$

The readers should compare this conclusion with Eq. 6.12.

Problems

6.1 For each case illustrated in Fig. 6.10, explain whether the situation is allowable or whether it is precluded by one or more laws of thermodynamics. The arrows indicate the direction of flow of work or heat, wavy lines being used for heat.

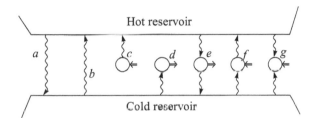

Figure 6.10

6.2 Prove that the two statements of the second law given in Section 6.2 are equivalent. Use negative logic as follows:

a. Assume that statement 1 is violated in some specific situation. This must be a heat engine with $\eta \geq 1$. Can the work produced by this engine be completely dissipated to thermal energy, which is transferred to the hot reservoir as heat? Can the work be used to drive an ordinary refrigerator operating between the reservoirs and thus augment the heat delivered to the hot reservoir? Would this combination of events constitutes a violation of the second statement in Section 6.2 of the text?

b. Assume that statement 2 is violated in some specific situation. This must be a case where heat flows spontaneously from a cold reservoir to a hot reservoir with no expenditure of work. Now install an ordinary heat engine to operate between these two same reservoirs. Adjust the heat flow to the cold reservoir to balance exactly the heat transferred between the reservoirs by the spontaneous process. Do the real heat engine and the spontaneous process together constitute a system that violates the first statement in Section 6.2?

    c. As an exercise in logic, do parts a and b of this problem establish the equivalence of the two statements in Section 6.2 of the text?

6.3 Let an ideal gas be a substance for which

$$p\tilde{V} = R\theta$$

and possibly

$$\left(\frac{\partial U}{\partial V}\right)_\theta = 0$$

where θ is the ideal-gas temperature. Use an ideal gas as the working substance in a Carnot engine operating between reservoirs at temperatures θ_H and θ_C. Calculate the efficiency in terms of θ_H and θ_C. Is the ideal-gas temperature the same as the thermodynamic temperature? Do you need to assume that the heat capacity

$$\tilde{C}_V = \left(\frac{\partial \tilde{U}}{\partial \theta}\right)_{\tilde{V}}$$

is independent of θ and $\tilde{V}$? Do you need to assume that $(\partial U/\partial V)_\theta$ is zero?

6.4 Assume that because of the limitations of the materials involved, the upper temperatures in a power plant are limited to 1150°F and that heat can be rejected to the environment at a temperature no lower than 90°F. Show on the basis of the efficiency of a Carnot engine that at least 5.47 MJ of heat must be supplied at the upper temperature in order to produce 1 kWh of work. (The more practical power cycle in Chapter 12 requires 8.51 MJ of heat to produce 1 kWh of work.)

Notation

COP	coefficient of performance for a refrigerator
n	number of moles
p	pressure, N/m^2
Q	heat added to system, J
R	universal gas constant, 8.3143 J/mol-K or 82.06 atm-cm^3/mol-K
S	entropy, J/K
T	thermodynamic temperature, K

U	internal energy, J
V	volume, m^3
W	work done by the system, J
η	efficiency
θ	arbitrary temperature scale
Φ	electric potential, V
ψ	thermometric function for a thermocouple, V

Subscripts

C	cold
H	hot

Chapter 7

Thermodynamic Relationships for PVT Systems

What do we need to measure in order to determine all the thermodynamic properties of a substance? How do we carry out the calculations from experimental result to tabulated quantity? The mathematical development of the present chapter is directed toward answering these questions. This is part of the effort, mentioned in Chapter 1, of making the program of property measurement as economical and effective as possible.

These mathematical exercises are worthwhile because a knowledge of the thermodynamic properties of substances permits one to make useful predictions about the behavior of matter in practical situations.

7.1   Volume Dependence of Internal Energy

For independent variables, let us at first select the volume V and the temperature T, the thermodynamic temperature scale being based on the Carnot cycle. We can measure experimentally the pressure p and the heat capacity C_V as functions of volume and temperature. These two state properties might be supposed to be independent, but we shall show that C_V need be known as a function of temperature

The Newman Lectures on Thermodynamics
John Newman and Vincent Battaglia
Copyright © 2019 Jenny Stanford Publishing Pte. Ltd.
ISBN 978-981-4774-26-0 (Hardcover), 978-1-315-10861-2 (eBook)
www.jennystanford.com

for only one volume V. This is the reason why the heat capacity in the ideal-gas limit was emphasized in Chapter 5, while the pressure was treated as a function of both temperature and volume in Chapter 3.

Again on the basis of the Carnot cycle, one can infer an entropy function S for which $dQ_{rev} = TdS$. Then, the first law of thermodynamics can be expressed for a reversible process as (see Eq. 6.27)

$$dU = dQ - dW = TdS - pdV. \tag{7.1}$$

Since U, S, and V are state properties, one can generally apply this equation to a reversible process between two states even though the physical process that led from one state to the other might be irreversible.

It follows from Eq. 7.1 that

$$p = -(\partial U/\partial V)_S \quad \text{and} \quad T = (\partial U/\partial S)_V. \tag{7.2}$$

Furthermore, since the second cross derivative is independent of the order of differentiation

$$\frac{\partial^2 U}{\partial S \partial V} = \frac{\partial^2 U}{\partial V \partial S}, \tag{7.3}$$

we conclude that

$$-\left(\frac{\partial p}{\partial S}\right)_V = \left(\frac{\partial T}{\partial V}\right)_S. \tag{7.4}$$

A result, like Eq. 7.4, which is obtained by the cross-differentiation of a state property, is known as a *Maxwell relation*.

The differential of the entropy can be written as

$$dS = \left(\frac{\partial S}{\partial V}\right)_T dV + \left(\frac{\partial S}{\partial T}\right)_V dT. \tag{7.5}$$

For variations at constant entropy, this yields

$$\left(\frac{\partial S}{\partial V}\right)_T = -\left(\frac{\partial S}{\partial T}\right)_V \left(\frac{\partial T}{\partial V}\right)_S \tag{7.6}$$

(compare Eq. 1.5). Substitution of Eq. 7.4 gives

$$\left(\frac{\partial S}{\partial V}\right)_T = \left(\frac{\partial S}{\partial T}\right)_V \left(\frac{\partial p}{\partial S}\right)_V = \left(\frac{\partial p}{\partial T}\right)_V. \tag{7.7}$$

Equation 7.7 allows the volume dependence of the entropy to be calculated solely from the PVT data. On the other hand, the definition of the heat capacity at constant volume

$$C_V = \left(\frac{dQ}{dT}\right)_V = T\left(\frac{\partial S}{\partial T}\right)_V \qquad (7.8)$$

provides us with the means to calculate the temperature dependence of the entropy. Equation 7.5 now becomes

$$\boxed{dS = \left(\frac{\partial p}{\partial T}\right)_V dV + \frac{C_V}{T}dT}. \qquad (7.9)$$

From the cross derivative of the entropy, we obtain

$$\boxed{\left(\frac{\partial C_V}{\partial V}\right)_T = T\left(\frac{\partial^2 p}{\partial T^2}\right)_V}, \qquad (7.10)$$

an important result that allows the heat capacity at a given volume to be calculated from the ideal-gas limit, along with the PVT data.

Substitution of Eq. 7.9 into Eq. 7.1 yields

$$dU = C_V dT + \left[T\left(\frac{\partial p}{\partial T}\right)_V - p\right]dV, \qquad (7.11)$$

from which we infer that

$$C_V = \left(\frac{\partial U}{\partial T}\right)_V \qquad (7.12)$$

(in agreement with Eq. 4.13) and

$$\boxed{\left(\frac{\partial U}{\partial V}\right)_T = T\left(\frac{\partial p}{\partial T}\right)_V - p}. \qquad (7.13)$$

The principal results obtained here are Eq. 7.9, from which variations in entropy can be calculated from PVT and heat-capacity data, and Eqs. 7.10 and 7.13, from which the volume dependence of heat capacity and internal energy can be determined from the PVT data.

7.2 Auxiliary Functions and Maxwell Relations

The above development indicates clearly and completely how differences in internal energy U and entropy S can be computed with the aid of experimental data on pressure as a function of volume and temperature and heat capacity C_V as a function of temperature at a given volume V. Several auxiliary functions are frequently defined. For PVT systems, these are the Helmholtz free energy

$$A = U - TS, \tag{7.14}$$

the enthalpy

$$H = U + pV, \tag{7.15}$$

and the Gibbs free energy or Gibbs function

$$G = U + pV - TS = H - TS = A + pV. \tag{7.16}$$

The Helmholtz free energy A is related to the maximum (reversible) work that can be extracted from a system (see Chapter 12) and also finds extensive application in statistical mechanics, the field dealing with the relationship between thermodynamic properties and molecular properties. From Eq. 7.1, the differential of A can be written immediately as

$$dA = -SdT - pdV. \tag{7.17}$$

The enthalpy H is related to the reversible heat added to a system which must expand against the constant pressure of the atmosphere (see Chapter 4). It also fits nicely into the energy balance (Eq. 4.34) for flow processes, where the shaft work is distinguished from the work related to the expansion or compression of the fluid. The enthalpy is used more often than the internal energy in the tables of thermodynamic data; for example, the steam tables [1] give values of volume, entropy, and enthalpy as functions of temperature and pressure. The differential of enthalpy is

$$dH = TdS + Vdp. \tag{7.18}$$

The enthalpy is sometimes regarded as a "heat function." This is a misleading notion, first of all, because heat is not a property of a system. The entropy might more accurately be regarded as a "heat function" because for any reversible process $đQ = TdS$. Thus, in Eq. 7.8, the heat capacity C_V was related directly to the entropy, and its relationship to the internal energy followed in Eq. 7.12. Similarly, the heat capacity at constant pressure is

$$C_p = \left(\frac{đQ}{dT} \right)_p = T \left(\frac{\partial S}{\partial T} \right)_p. \tag{7.19}$$

From Eq. 7.18, it now follows that

$$C_p = \left(\frac{\partial H}{\partial T} \right)_p. \tag{7.20}$$

Enthalpy as a "heat function" is misleading, secondly, because it is specific to PVT systems. An electrochemical cell, for example a lead–acid battery for your automobile, undergoing isothermal discharge at constant atmospheric pressure absorbs an amount of heat equal to $T\Delta S$, not ΔH. The enthalpy change will be negative (about −1.64 J/C), while $T\Delta S$ may be positive for some battery systems and negative for others and is probably much smaller in magnitude than ΔH (for example, $T\Delta S$ is about +0.40 J/C for the lead–acid battery).

The Gibbs function G, whose differential is

$$dG = -SdT + Vdp, \tag{7.21}$$

is useful for describing internal equilibrium in a system. For example, the change in the Gibbs function is zero for the reversible transformation of liquid water into steam (see Chapter 11). This means that the Gibbs free energy per unit mass, $\hat{G}$, is the same for these two phases in equilibrium.

From the simple form of the differentials in Eqs. 7.1, 7.17, 7.18, and 7.21, it might appear that the logical choice of independent variables would be those given in Table 7.1. However, any two of the variables can generally be selected as the independent variables. From these equations for the differentials, there follow the relations in Eq. 7.2 as well as

$$p = -\left(\frac{\partial A}{\partial V}\right)_T, \quad T = -\left(\frac{\partial H}{\partial S}\right)_p, \tag{7.22}$$

$$S = -\left(\frac{\partial A}{\partial T}\right)_V = -\left(\frac{\partial G}{\partial T}\right)_p, \tag{7.23}$$

and

$$V = \left(\frac{\partial H}{\partial p}\right)_S = \left(\frac{\partial G}{\partial p}\right)_T. \tag{7.24}$$

Table 7.1 Independent variables which arise in a natural way when using the thermodynamic energy functions U, H, A, and G.

Function	"Logical" independent variables
U	S, V
H	S, p
A	T, V
G	T, p

However, it makes just as much sense to observe that

$$\frac{p}{T} = \left(\frac{\partial S}{\partial V}\right)_U \quad \text{or} \quad \frac{S}{V} = \left(\frac{\partial p}{\partial T}\right)_G, \tag{7.25}$$

and the proliferation of equations serves no useful purpose at this point. (The first of Eq. 7.22 does find application in statistical mechanics.)

Equations 7.17, 7.18, and 7.21 yield the Maxwell relations

$$\left(\frac{\partial S}{\partial V}\right)_T = \left(\frac{\partial p}{\partial T}\right)_V, \tag{7.26}$$

$$\left(\frac{\partial T}{\partial p}\right)_S = \left(\frac{\partial V}{\partial S}\right)_p, \tag{7.27}$$

and

$$-\left(\frac{\partial S}{\partial p}\right)_T = \left(\frac{\partial V}{\partial T}\right)_p. \tag{7.28}$$

Equation 7.26 had already been derived by another method (see Eq. 7.7 and Problem 7.1). Equation 7.28 may be more practical than Eq. 7.26 for obtaining values of the entropy because it involves PVT experiments at constant pressure instead of constant volume.

7.3   Relationship of Heat Capacities

Equations 7.19 and 7.28 permit the differential of S to be expressed as

$$\boxed{dS = \frac{C_p}{T} dT - \left(\frac{\partial V}{\partial T}\right)_p dp}. \tag{7.29}$$

This equation, which is similar to Eq. 7.9, shows how to calculate entropy changes with temperature and pressure on the basis of heat capacity and PVT data. From Eq. 7.9,

$$\left(\frac{\partial S}{\partial T}\right)_p = \frac{C_V}{T} + \left(\frac{\partial p}{\partial T}\right)_V \left(\frac{\partial V}{\partial T}\right)_p. \tag{7.30}$$

Comparison with Eq. 7.29 or Eq. 7.19 shows that

$$\boxed{C_p = C_V + T\left(\frac{\partial p}{\partial T}\right)_V \left(\frac{\partial V}{\partial T}\right)_p}. \tag{7.31}$$

On the basis of Eq. 1.5, this relationship between the heat capacities at constant volume and at constant pressure can also be expressed as

$$C_p = C_V + TV\beta^2/k, \tag{7.32}$$

that is, in terms of the coefficient of thermal expansion β and the isothermal compressibility k.

The second cross-derivative of entropy with respect to temperature and pressure yields (from Eq. 7.29) the pressure dependence of C_p:

$$\boxed{\left(\frac{\partial C_p}{\partial p}\right)_T = -T\left(\frac{\partial^2 V}{\partial T^2}\right)_p} \tag{7.33}$$

(compare Eq. 7.10). Finally, substitution of Eq. 7.29 into Eq. 7.18 yields

$$dH = C_p dT + \left[V - T\left(\frac{\partial V}{\partial T}\right)_p\right]dp, \tag{7.34}$$

from which we deduce the pressure dependence of the enthalpy:

$$\boxed{\left(\frac{\partial H}{\partial p}\right)_T = V - T\left(\frac{\partial V}{\partial T}\right)_p} \tag{7.35}$$

Example 7.1 Calculate the percent difference between C_V and C_p for an ideal diatomic gas and for water, for which the compressibility and the coefficient of thermal expansion at 20°C are

$$k = -\frac{1}{V}\left(\frac{\partial V}{\partial p}\right)_T = 4.58\times 10^{-11}\frac{\text{cm}^2}{\text{dyne}}$$

and

$$\beta = \frac{1}{V}\left(\frac{\partial V}{\partial T}\right)_p = 0.207\times 10^{-3}\text{K}^{-1},$$

the heat capacity is $\hat{C}_p = 0.99883$ cal/g-K = 4.1819 J/g-K, and the density is $\rho = 0.99823$ g/cm³.

Solution: For the ideal diatomic gas, take $\tilde{C}_V = 5R/2$ and $\tilde{C}_p = 7R/2$. Then

$$\frac{\tilde{C}_p - \tilde{C}_V}{\tilde{C}_V} = \frac{R}{5R/2} = 0.4 = 40\%.$$

For water, we use Eq. 7.32.

$$\hat{C}_p - \hat{C}_V = \frac{293.15 \times (0.207 \times 10^{-3})^2 \times 10^{-7}}{0.99823 \times 4.58 \times 10^{-11}} = 0.0275 \frac{J}{g-K}.$$

Now

$$\frac{\hat{C}_p - \hat{C}_V}{\hat{C}_V} = \frac{0.0275}{4.1819 - 0.0275} = 0.66\%.$$

Example 7.2 A process at a constant temperature of 25°C and a constant pressure of 1 atm involves the reversible discharge of the cell

$$\xleftarrow{\hspace{1cm}} -0.35232 \text{ V} + \xrightarrow{\hspace{1cm}}$$

$$\text{Pt, H}_2 \left| \begin{array}{c} 0.1 \text{ mol HCl} \\ \text{kg H}_2\text{O} \end{array} \right| \text{AgCl} \left| \text{Ag} \right| \text{Pt}$$

(see Fig. 7.1) according to the discharge reaction

$$\frac{1}{2} \text{ H}_2 + \text{AgCl} \rightarrow \text{HCl} + \text{Ag}.$$

Calculate ΔG, ΔH, and $p\Delta V$ from the reversible cell potential and the value of $T\Delta S = -0.05489$ J/C transferred.

Solution: In this example, we use as a basis the transfer of 1 C of charge through the cell and through the external circuit. One joule per coulomb is the same as a volt. The law of conservation of energy says

$$\Delta U = Q - W = T\Delta S - p\Delta V - 0.35232 \text{ J/C}.$$

Hence, in view of the constant temperature and pressure,

$$\Delta H = \Delta(U + pV) = -0.05489 - 0.35232 = -0.40722 \text{ J/C}$$

and

$$\Delta G = \Delta(U + pV - TS) = -0.35232 \text{ J/C}.$$

Thus, the change in the Gibbs function (per coulomb transferred) is the same as the cell potential, but with the opposite sign.

The overall cell reaction can be regarded to be composed of the anode reaction

$$\frac{1}{2} \text{H}_2 \rightarrow \text{H}^+ + e^-$$

and the cathode reaction

$$e^- + AgCl \rightarrow Ag + Cl^-.$$

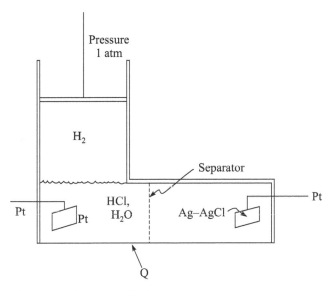

Figure 7.1 Electrochemical cell with a hydrogen electrode to the left and a silver–silver chloride electrode to the right.

The consumption of half a mole of hydrogen is accompanied by the transfer of 1 mol of electrons carrying a charge of 1 F or 96,487 C. With the ideal-gas law for hydrogen, the term $p\Delta V$ amounts to

$$p\Delta V = -\frac{RT}{2F} = -\frac{8.3143 \times 298.15}{2 \times 96,487} = -0.0128 \text{ J/C}.$$

(For electrochemical cells that do not involve the consumption or production of a gas phase, the $p\Delta V$ term will be much smaller.)

Notice that ΔH is seven times larger than $Q = T\Delta S$.

Example 7.3 For water at 20°C, calculate $(\partial \tilde{U} / \partial p)_T$ using the data from Example 7.1.

Solution: From Eqs. 7.13 and 1.5, we obtain

$$\left(\frac{\partial \tilde{U}}{\partial p}\right)_T = \left(\frac{\partial \tilde{U}}{\partial V}\right)_T \left(\frac{\partial V}{\partial p}\right)_T = -\tilde{V}[T\beta - pk].$$

With $\tilde{V}$ = 18.015/0.99823 = 18.047 cm³/mol and p = 1 atm = 1.013×10⁶ dyne/cm², we obtain

$$\left(\frac{\partial \tilde{U}}{\partial p}\right)_T = -\tilde{V}\,[293.15 \times 0.207 \times 10^{-3} - 4.58 \times 10^{-11} \times 10^6]$$

$$= -\tilde{V}\,[0.06068 - 0.00005] = -0.06063\,\tilde{V} = -1.0942\ \text{cm}^3/\text{mol}.$$

Notice that the contribution of the term pk is negligible at a pressure of 1 atm.

Example 7.4 Develop the thermodynamic properties of a van der Waals fluid for which the heat capacity in the ideal-gas state is $\tilde{C}_V^* = 5R/2$, independent of temperature.

Solution: We take the parameters a and b to be fitted to the critical properties according to Eqs. 3.11, and the critical volume is given by Eq. 3.12. Reduced properties are defined as follows:

$$p_r = \frac{p}{p_c},\ T_r = \frac{T}{T_c},\ V_r = \frac{\tilde{V}}{\tilde{V}_c},$$

$$H_r = \frac{\tilde{H}}{RT_c},\ S_r = \frac{\tilde{S}}{R},\ G_r = \frac{\tilde{G}}{RT_c},\ U_r = \frac{\tilde{U}}{RT_c}.$$

From the equation of state 3.10, the reduced pressure is

$$p_r = \frac{8T_r}{3V_r - 1} - \frac{3}{V_r^2},$$

and the variations with volume and temperature are

$$\left(\frac{\partial p}{\partial \tilde{V}}\right)_T = -\frac{p_c}{\tilde{V}_c}\left[\frac{24T_r}{(3V_r - 1)^2} - \frac{6}{V_r^3}\right]$$

and

$$\left(\frac{\partial p}{\partial T}\right)_V = \left(\frac{\partial S}{\partial V}\right)_T = \frac{R}{\tilde{V} - b}.$$

Integration for the entropy gives

$$\tilde{S} = R\ln\left(\frac{\tilde{V} - b}{2b}\right) + f(T),$$

and differentiation with respect to temperature yields

$$\left(\frac{\partial \tilde{S}}{\partial T}\right)_{\tilde{V}} = f'(T) = \frac{\tilde{C}_V}{T}.$$

In order for $\tilde{C}_V$ to approach the ideal-gas value, $5R/2$, as $\tilde{V}$ approaches infinity, we must have

$$f = \frac{5}{2}R\ln\frac{T}{T_c} + \text{constant}.$$

When the constant is chosen so that the entropy is zero at the critical point, the reduced entropy can be written as

$$S_r = \ln\left(\frac{3V_r - 1}{2}\right) + \frac{5}{2}\ln T_r.$$

Note that for a van der Waals fluid, $\tilde{C}_V$ is independent of volume and depends only on temperature.

From Eq. 7.13, the volume dependence of the internal energy is

$$\left(\frac{\partial U}{\partial V}\right)_T = \frac{a}{\tilde{V}^2} = \frac{9RT_c\tilde{V}_c}{8\tilde{V}^2}.$$

Integration gives

$$\tilde{U} = F(T) - \frac{9RT_c\tilde{V}_c}{8\tilde{V}}.$$

Selection of $F(T)$ so that $(\partial\tilde{U}/\partial T)_V = \tilde{C}_V$ allows us to express the reduced internal energy as

$$U_r = \frac{5}{2}T_r - \frac{9}{8V_r} - \frac{7}{4},$$

the constant being selected so that the enthalpy is zero at the critical point (see below).

It is now a simple matter to write down the reduced enthalpy

$$H_r = U_r + Z_c p_r V_r = \frac{7}{2}T_r + \frac{T_r}{3V_r - 1} - \frac{9}{4V_r} - \frac{7}{4}$$

and the reduced Gibbs function is

$$G_r = H_r - T_r S_r = \frac{7}{2}T_r - \frac{5}{2}T_r\ln\left(\frac{3V_r - 1}{2}\right) + \frac{T_r}{3V_r - 1} - \frac{9}{4V_r} - \frac{7}{4}.$$

Example 7.5 Two masses of nitrogen, which can be treated as an ideal gas, are mixed irreversibly under the conditions of Example

5.3 by the removal of the separating diaphragm. The heat capacity of nitrogen can be taken to be constant and given by $\tilde{C}_p^* = 7R/2$. Calculate the entropy change of the system. Is the entropy change of the universe the same, greater than, or less than the value calculated for the system?

Solution: Application of Eq. 7.28 to an ideal gas yields

$$\left(\frac{\partial \tilde{s}}{\partial p}\right)_T = -\left(\frac{\partial \tilde{v}}{\partial T}\right)_p = -\frac{R}{p},$$

and integration gives

$$\tilde{s} = -R \ln p + f(T).$$

Differentiation with respect to T gives

$$\left(\frac{\partial \tilde{s}}{\partial T}\right)_p = f'(T) = \frac{\tilde{C}_p}{T}.$$

Consequently, $f(T) = \tilde{C}_p \ln T + \text{constant}$, and the molar entropy of an ideal gas is

$$\tilde{s} = \tilde{C}_p \ln T - R\ln p + \text{constant}.$$

Example 5.3 gives the initial numbers of moles and the final temperature and pressure of the gas. Taking as a basis 1 mol of the final mixture, we express the entropy change for the system as

$$\frac{\Delta S}{n} = \tilde{C}_p \ln T_f - R \ln p_f - \frac{n_1}{n}\left[\tilde{C}_p \ln T_1 - R \ln p_1\right] - \frac{n_2}{n}\left[\tilde{C}_p \ln T_2 - R \ln p_2\right].$$

Substitution of numerical values from Example 5.3 gives

$$\frac{\Delta S}{nR} = \frac{n_1}{n}\left[\frac{7}{2}\ln \frac{T_f}{T_1} - \ln \frac{p_f}{p_1}\right] + \frac{n_2}{n}\left[\frac{7}{2}\ln \frac{T_f}{T_2} - \ln \frac{p_f}{p_2}\right]$$

$$= 0.04575 + 0.000097 = 0.04585.$$

The larger of these numerical contributions is from the pressure terms. The entropy change of the system thus is $\Delta S/n$ = 0.0911 Btu/lb-mol-°R. The entropy change of the universe is the same as that calculated for the system, since the system is isolated.

Problems

7.1 The Maxwell relation 7.7 was derived from Eq. 7.4 without defining the Helmholtz free energy A. In a similar manner, obtain the Maxwell relations 7.27 and 7.28 without using explicitly the differentials of H and G.

7.2 Show that the pressure dependence of the internal energy can be expressed as

$$\left(\frac{\partial U}{\partial p}\right)_T = -T\left(\frac{\partial V}{\partial T}\right)_p - p\left(\frac{\partial V}{\partial p}\right)_T = -V(T\beta - pk)$$

and that the volume dependence of the enthalpy can be expressed as

$$\left(\frac{\partial H}{\partial V}\right)_T = T\left(\frac{\partial p}{\partial T}\right)_V + V\left(\frac{\partial p}{\partial V}\right)_T = \frac{T\beta - 1}{k}.$$

7.3 Show that the expression for the reduced Gibbs function obtained in Example 7.4 is in agreement with the equations

$$\left(\frac{\partial G}{\partial p}\right)_T = V \text{ and } \left(\frac{\partial G}{\partial T}\right)_p = -S.$$

7.4 Develop the thermodynamic properties of a Redlich–Kwong fluid for which the heat capacity in the ideal-gas state is $\tilde{C}_V^* = 5R/2$, independent of temperature.

7.5 In convective heat transfer, the variation of temperature for a pure fluid can be expressed as

$$C_V dT = đQ - T\left(\frac{\partial p}{\partial T}\right)_V dV.$$

For a condensed phase, the density is relatively constant, and we might like to write $dV = 0$. However, the coefficient $(\partial p / \partial T)_V$ is large for a condensed phase. To avoid this problem of multiplying a small number by a large number, obtain the completely equivalent equation

$$\hat{C}_p dT = đQ + T\left(\frac{\partial \hat{V}}{\partial T}\right)_p dp.$$

For water at 20°C, a change in pressure of 1 atm in the last term corresponds to what temperature change in the first term? (See Example 7.1 for the necessary data for water.)

7.6 Careful measurements on a pure substance show that the heat capacity $\tilde{C}_V^*$ is equal to 2.5R, independent of both temperature and pressure. What is the most general pVT equation of state that can apply to this substance, subject to the constraint that the ideal-gas law is obeyed at low pressures?

7.7 In 1914, Wohl proposed the following pVT equation of state:

$$p = \frac{RT}{\tilde{V} - b} - \frac{a}{\tilde{V}(\tilde{V} - b)} + \frac{c}{\tilde{V}^3}.$$

The constants a, b, and c can be expressed in terms of critical properties as follows:

$$a = 6\tilde{V}_c^2 p_c, \quad b = \frac{1}{4}\tilde{V}_c, \quad \text{and} \quad c = 4\tilde{V}_c^3 p_c,$$

where the molar volume at the critical point can itself be expressed in terms of the critical temperature and pressure:

$$\tilde{V}_c = \frac{RT_c}{3.75 p_c}.$$

Show that this equation of state approaches the ideal-gas law at low pressures.

If the ideal-gas heat capacity $\tilde{C}_V^*$ is a constant equal to 2.5R, obtain expressions for the temperature and volume dependence of the molar entropy $\tilde{S}_i(T, \tilde{V})$ and of the heat capacity $\tilde{C}_V(T, \tilde{V})$.

7.8 Show that the variation of temperature with pressure for a reversible, adiabatic compression or expansion (that is, for an isentropic process) is given by

$$\left(\frac{\partial T}{\partial p}\right)_S = \frac{T}{C_p}\left(\frac{\partial V}{\partial T}\right)_p.$$

Notation

a	van der Waals constant, atm-cm^6/mol^2
A	Helmholtz free energy, J
b	van der Waals constant, cm^3/mol
C_p	heat capacity at constant pressure, J/K
C_V	heat capacity at constant volume, J/K
$f(T)$	integration constant, a function of temperature, J/mol-K

F Faraday's constant, 96,487 C/mol

$F(T)$ integration constant, a function of temperature, J/mol

G Gibbs function, J

H enthalpy, J

k isothermal compressibility, m^2/N

n number of moles

p pressure, N/m^2

Q heat added to the system, J

R universal gas constant, 8.3143 J/mol-K or
82.06 atm-cm^3/mol-K

S entropy, J/K

T thermodynamic temperature, K

U internal energy, J

V volume, m^3

W work done by the system, J

Z compressibility factor

β coefficient of thermal expansion, K^{-1}

ρ density, g/cm^3

Subscripts, superscripts, and special symbols

c critical point

r reduced property

rev   reversible

$*$ ideal-gas state

$\sim$ per mole

$\wedge$ per unit mass

Reference

1. R. W. Bain. *Steam Tables 1964. Physical Properties of Water and Steam, 0–800°C, 0–1000 bars*. Edinburgh: Her Majesty's Stationary Office, 1964.

Chapter 8

Entropy, Irreversibility, Randomness, and Natural Philosophy

What is the physical meaning of entropy? The introduction of entropy may seem somewhat abstract, and some discussion of its properties should help to put it into perspective. Three points deserve emphasis:

1. In an *irreversible* process, the entropy change of the universe is always positive, while in a reversible process, the entropy change of the universe is zero. The entropy of the universe *never* decreases.

2. From a molecular point of view, an increase in entropy corresponds to an increase in randomness.

3. From a practical point of view, the reversible heat transfer is directly related to the entropy change, $dQ_{\text{rev}} = TdS$.

To demonstrate the first of these statements, we should first develop the concept of *lost work*. During a process, the system interacts with the surroundings in going from the initial state to the final state (see Fig. 8.1a). The *lost work* W_L for the process is defined as the work that must be extracted from an independent, reversible work source in order to restore the system *and* its surroundings to their initial conditions (Fig. 8.1b). During this restoration process, which should be reversible, heat can be rejected to an independent,

The Newman Lectures on Thermodynamics
John Newman and Vincent Battaglia
Copyright © 2019 Jenny Stanford Publishing Pte. Ltd.
ISBN 978-981-4774-26-0 (Hardcover), 978-1-315-10861-2 (eBook)
www.jennystanford.com

reversible heat source at the temperature T_0. Let us relate W_L to the entropy change of the "universe" during the forward process.

During the forward process, the system and its surroundings, which constitute the "universe," are isolated. Therefore,

$$\Delta U_{\text{universe}} = 0, \tag{8.1}$$

and at the same time the entropy changes by the amount $\Delta S_{\text{universe}}$.

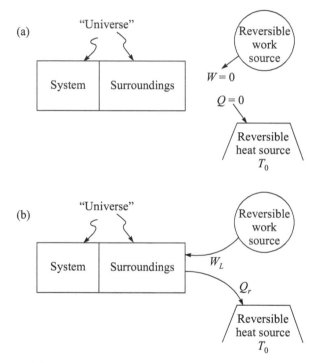

Figure 8.1 (a) Forward process, which may be irreversible; (b) restoration process, which should be reversible. For any process, a restoration is conceivable, in which the system and its surroundings are returned to their original states. However, this may require interaction with an external work source and an external heat source.

During the restoration process, the "universe" is returned to its original state, and its changes of internal energy and entropy are equal but opposite to those occurring in the forward process. Since $\Delta U_{\text{universe}}$ is zero, this means that

$$W_L = Q_r, \tag{8.2}$$

the work extracted from the work source is equal to the heat rejected to the heat source.

For the reversible restoration process, the total entropy change is zero, and this is the sum of the entropy change, now $-\Delta S_{universe}$, for the system and its surroundings and the entropy change Q_r/T_0 for the reversible heat source:

$$-\Delta S_{universe} + Q_r/T_0 = 0, \qquad (8.3)$$

since the reversible work source has no entropy change. Substitution for Q_r gives the final result

$$W_L = T_0 \Delta S_{universe}. \qquad (8.4)$$

The lost work is thus directly proportional to the entropy change of the universe. For a reversible process, then, W_L is zero, and we have a macroscopic definition of a reversible process: The system and its surroundings can be restored to their original states after a reversible process without the intervention of work from an external agency. Either W_L or $\Delta S_{universe}$ is a quantitative measure of the irreversibility of a process. The lost work has the advantage of appealing to our physical intuition. The entropy change of the universe has the advantage of being independent of the temperature T_0, which is, after all, extraneous to the forward process (see Fig. 8.1).

Let us return to our statements of the second law of thermodynamics at the end of Chapter 6. In developing the relationship of the lost work to the entropy change of the universe, we have used statement 4, that entropy is a state property having the significance $dQ_{rev} = TdS$. From Eq. 8.4, it is now apparent that statements 2 and 5 are equivalent. Otherwise, if $\Delta S_{universe}$ were negative, we should have, with the restoration process, a device for extracting the heat $-Q_r$ from a reservoir and completely converting it into the work $-W_L$, without making any other change in the universe.

In statistical mechanics, or from a molecular point of view, an increase in entropy corresponds to an increase in randomness:

$$S = k \ln \Omega, \qquad (8.5)$$

where Ω is the number of microscopic states that can correspond to a given macroscopic state and k is the Boltzmann constant. We can perhaps appreciate the fact that we have a more ordered situation in Examples 5.3 and 7.5 before we remove the diaphragm and allow the gases to mix. Problem 8.4 is, however, an example in which the number of microscopic states does not correspond to geometric order.

The principle of entropy has important philosophical implications. The fact that the entropy of the universe is always increasing and that a system with a higher entropy is more random suggests that the entire universe is approaching an equilibrium state, one filled uniformly with dilute, grey matter devoid of life, form, and temperature differences. This can influence considerably the thinking of prophets of doom.

We should point out that no cosmic significance was intended for the word *universe*; it was merely a convenient, simple word for the thermodynamic system and the immediate surroundings with which the system interacts. In Fig. 8.1, we even had something outside the "universe." Strictly, the second law of thermodynamics applies to a restricted realm, and philosophical conclusions are not without doubt.

Natural science has also exerted a profound influence on the philosophical question of free agency and determinism. Newtonian mechanics implies that the positions and velocities of all bodies can be calculated at any time if they are known at some initial time. If this conclusion applies to molecules and subatomic particles, the implication is that the future is entirely determined by the present and that, therefore, people have no ability to make any choices for themselves. This philosophical conclusion is tenuous by itself, and the advent of quantum mechanics has weakened it more. The Heisenberg uncertainty principle says that we cannot know with complete precision the initial positions and velocities of all particles. The readers are free to develop their own ideas on these philosophical matters.

Problems

8.1 As an engineer in a small chemical plant, you have available to you a heat source of 100,000 Btu at 400°F and another heat source of 100,000 Btu at 212°F. If power is worth 3.6 ¢/kWh and your ultimate heat sink is at 70°F, how much value should you attach to each of these energy sources?

8.2 Chapter 6 treated irreversible compression and expansion of a nearly isothermal gas. For this cyclic process, discuss the entropy changes for the system, the heat reservoir, the work source, and the universe.

8.3 Suppose that statement 4 at the end of Chapter 6 is valid, that is, there exists a state property S such that $dQ_{rev} = TdS$. For a PVT system, this means that S is known as a function of internal energy U and volume V.

    a. Show how to calculate the temperature T and the pressure p from the given function $S(U, V)$.

    b. Suppose that it is possible to construct an engine that, operating in a cycle, will produce no effect other than the extraction of heat from a reservoir and the performance of an equivalent amount of work. Justify that this must be a reversible process on the basis of the definition in the paragraph below Eq. 8.4.

    c. Calculate the entropy change of the engine on the basis of statement 4 that $dQ_{rev} = TdS$.

    d. If the engine operates in a cycle and entropy is a state property, what is the entropy change of the engine? Discuss, on this basis, how statement 2 at the end of Chapter 6 is a consequence of statement 4.

8.4 Suppose that we have a supersaturated, aqueous solution of sodium sulfate. Introduction of a minute crystal of sodium sulfate leads to the formation of large crystals and a solution in equilibrium with this solid.

If the system is otherwise isolated, the temperature rises during this process. Is the final state more ordered or more random than the initial state? Is the final entropy higher or lower than the initial entropy? Is a system with a higher entropy more random?

8.5 As the superintendent of a small chemical plant, you have to evaluate a proposal made by a recently graduated engineer. The lost work W_L for the plant is $T_0 \Delta S_{universe}$, where $\Delta S_{universe}$ is the entropy *production* within the plant. Without modifying the operation of the plant itself, the engineer proposes to reduce W_L. He wants to lower the value of T_0 by building a heat sink at a lower temperature, using refrigeration to achieve the lower temperature.

Notation

k Boltzmann constant, 1.38×10^{-23} J/K

p pressure, N/m^2

Q heat added to system, J

Q_r external heat transferred in restoration process, J

S entropy, J/K

T thermodynamic temperature, K

T_0 temperature of reversible heat source, K

U internal energy, J

V volume, m^3

W_L lost work, J

Ω number of microscopic states that can correspond to a given macroscopic state

Subscript

rev   reversible

Chapter 9

The Virial Equation

A gas satisfies the virial equation of state

$$p\tilde{V} = A(1 + B/\tilde{V} + C/\tilde{V}^2 + \cdots),\tag{9.1}$$

where A, B, C, etc. are independent of the molar volume $\tilde{V}$ but do depend on temperature as well as the nature of the particular gas. This equation embodies Boyle's law: at low pressures, the product of pressure and volume is a constant at a given temperature.

Substitution into Eq. 7.13 yields

$$\left(\frac{\partial \tilde{U}}{\partial \tilde{V}}\right)_T = \frac{T^2}{\tilde{V}}\left(\frac{dA/T}{dT} + \frac{1}{\tilde{V}}\frac{dAB/T}{dT} + \frac{1}{\tilde{V}^2}\frac{dAC/T}{dT} + \cdots\right).\tag{9.2}$$

Integration at constant temperature now gives

$$\tilde{U} = \tilde{U}^*(T) + T^2\frac{dA/T}{dT}\ln\tilde{V} - \frac{T^2}{\tilde{V}}\frac{dAB/T}{dT} - \frac{T^2}{2\tilde{V}^2}\frac{dAC/T}{dT} + \cdots.\tag{9.3}$$

Here, $\tilde{U}^*$ is the integration constant, which can depend on temperature.

Equation 9.3 shows that the internal energy approaches infinity at low pressures unless A is proportional to the temperature T:

$$A = RT.\tag{9.4}$$

Experimental results show that $\tilde{U}$ is, in fact, bounded at low pressures (Fig. 5.5 in Ref. [1]). The same conclusion is reached in

The Newman Lectures on Thermodynamics
John Newman and Vincent Battaglia
Copyright © 2019 Jenny Stanford Publishing Pte. Ltd.
ISBN 978-981-4774-26-0 (Hardcover), 978-1-315-10861-2 (eBook)
www.jennystanford.com

statistical mechanics. This molecular theory relates the internal energy to three contributions. The translational energy of the molecules is $3RT/2$ and is independent of pressure. The rotational and vibrational energies within the molecules should become independent of pressure at low pressures, and the potential energy of interaction between the molecules should approach zero at low pressures (see Fig. 3.4).

Thus, we are led to conclude that Eq. 9.4 applies. The arguments in Chapter 3 then show that the constant R is independent of the mass of the gas and that it is a universal constant, independent of the nature of the gas, by the definition of the mole. With Eq. 9.4, Eq. 9.1 becomes identical to Eq. 3.19, and Eqs. 9.2 and 9.3 reduce to

$$\left(\frac{\partial \widetilde{U}}{\partial \widetilde{V}}\right)_T = \frac{RT^2}{\widetilde{V}^2}\frac{dB}{dT} + \frac{RT^2}{\widetilde{V}^3}\frac{dC}{dT} + \cdots \tag{9.5}$$

and

$$\widetilde{U} = \widetilde{U}^*(T) - \frac{RT^2}{\widetilde{V}}\frac{dB}{dT} - \frac{RT^2}{2\widetilde{V}^2}\frac{dC}{dT} + \cdots . \tag{9.6}$$

$\widetilde{U}^*(T)$ is thus the internal energy in the ideal-gas state; its derivative is the ideal-gas limit of $\widetilde{C}_V$, discussed in Chapter 5.

Since Eq. 7.13 involves the thermodynamic temperature, T in Eq. 9.4 is also the thermodynamic temperature. We have thus demonstrated the equivalence of the ideal-gas and thermodynamic temperature scales, without taking either an ideal gas or a real gas through a Carnot cycle (see Section 6.3). We have also demonstrated that to the extent that the ideal-gas law, Eq. 3.8, applies, so also the internal energy is independent of volume (see Eq. 5.9). Finally, we should like to point out that both Charles's and Gay-Lussac's laws have been derived from Boyle's law.

For the pressure dependence of the internal energy, we have

$$\left(\frac{\partial \widetilde{U}}{\partial p}\right)_T = \left(\frac{\partial \widetilde{U}}{\partial \widetilde{V}}\right)_T\left(\frac{\partial \widetilde{V}}{\partial p}\right)_T = -T\frac{\dfrac{dB}{dT} + \dfrac{1}{\widetilde{V}}\dfrac{dC}{dT} + \cdots}{1 + \dfrac{2B}{\widetilde{V}} + \dfrac{3C}{\widetilde{V}^2} + \cdots}. \tag{9.7}$$

We should note that this derivative does not approach zero at zero pressure (see Example 9.1). In this respect, the ideal gas is not a good approximation to the properties of real gases.

The volume dependence of the heat capacity is developed in Example 9.2.

With the virial Eq. 3.21,

$$p\tilde{V} / RT = 1 + B'p + C'p^2 + \cdots, \tag{9.8}$$

where the compressibility factor is expanded in terms of the pressure, it might be more convenient to work with the enthalpy and the pressure. Equation 7.33 yields

$$\left(\frac{\partial \tilde{C}_p}{\partial p}\right)_T = -RT\left[\frac{d^2 B'T}{dT^2} + p\frac{d^2 C'T}{dT^2} + \cdots\right], \tag{9.9}$$

and integration at constant temperature gives

$$\tilde{C}_p = \tilde{C}_p^*(T) - pRT\frac{d^2 B'T}{dT^2} - \frac{p^2 RT}{2}\frac{d^2 C'T}{dT^2} - \cdots, \tag{9.10}$$

while Eq. 7.35 becomes

$$\left(\frac{\partial \tilde{H}}{\partial p}\right)_T = -T^2\left(\frac{\partial \tilde{V}/T}{\partial T}\right)_p = -RT^2\left[\frac{dB'}{dT} + p\frac{dC'}{dT} + \cdots\right]. \tag{9.11}$$

Example 9.1 Estimate $(\partial \tilde{U} / \partial p)_T$ for carbon dioxide at 1 atm and 20°C.

Solution: At low pressures, the pressure dependence of the internal energy can be related to the second virial coefficient:

$$\left(\frac{\partial \tilde{U}}{\partial p}\right)_T = -T\frac{dB}{dT}.$$

From the corresponding-states correlation of the second virial coefficient (Eq. 3.23), we obtain

$$\left(\frac{\partial \tilde{U}}{\partial p}\right)_T = -\tilde{V}_c\left[0.881\frac{T_c}{T} + 2\times 0.757\left(\frac{T_c}{T}\right)^2\right].$$

From the critical properties in Table 3.1, the reduced temperature is

$$T / T_c = 293.15 / 304 = 0.9643$$

and the critical volume is

$$\tilde{V}_c = \frac{Z_c RT_c}{p_c} = \frac{0.276 \times 304 \times 82.054}{72.9} = 94.44 \text{ cm}^3/\text{mol}.$$

Substitution into the above equation gives

$$\left(\frac{\partial \tilde{U}}{\partial p}\right)_T = -2.542\,\tilde{V}_c = -240 \text{ cm}^3/\text{mol}.$$

Equation 9.5 shows that $(\partial \tilde{U}/\partial \tilde{V})_T$ approaches zero at zero pressure for real gases. However, $(\partial \tilde{U}/\partial p)_T$ does not approach zero; it is zero only for "ideal gases," those which obey Eq. 3.8 exactly. The present example allows us to estimate $(\partial \tilde{U}/\partial p)_T$ for real gases. Comparison with the result of Example 7.3 shows that $(\partial \tilde{U}/\partial p)_T$ is, in fact, much larger for gases than for liquids.

Example 9.2 Estimate the heat capacity $\tilde{C}_V$ for saturated water vapor at 20°C. The ideal-gas value can be taken to be $\tilde{C}_V^* = 6.06$ cal/mol-K.

Solution: From Eqs. 7.10 and 3.19,

$$\left(\frac{\partial C_V}{\partial V}\right)_T = T\left(\frac{\partial^2 p}{\partial T^2}\right)_V = \frac{RT}{\tilde{V}^2}\left[\frac{d^2 BT}{dT^2} + \frac{1}{\tilde{V}}\frac{d^2 CT}{dT^2} + \frac{1}{\tilde{V}^2}\frac{d^2 DT}{dT^2} + \cdots\right].$$

Integration at constant temperature gives

$$\tilde{C}_V = \tilde{C}_V^*(T) - \frac{RT}{\tilde{V}}\left[\frac{d^2 BT}{dT^2} + \frac{1}{2\tilde{V}}\frac{d^2 CT}{dT^2} + \frac{1}{3\tilde{V}^2}\frac{d^2 DT}{dT^2} + \cdots\right].$$

For saturated water at 20°C, p = 0.3388 psia, and p/p_c = 0.3388/3206.2 = 1.057 × 10^{-4} and T/T_c = 293.15/647 = 0.453. Under these conditions, let us use the corresponding-states Eq. 3.23 to evaluate $d^2 BT/dT^2$, and let us neglect terms involving the higher-order virial coefficients.

$$\tilde{C}_V = \tilde{C}_V^* + \frac{R\tilde{V}_c}{\tilde{V}}1.514\left(\frac{T_c}{T}\right)^2 + O(\tilde{V}^{-2})$$

$$= \tilde{C}_V^* + 1.514\frac{Z_c}{Z}R\frac{p}{p_c}\left(\frac{T_c}{T}\right)^3 = \tilde{C}_V^* + R\frac{1.514\times 0.250\times 1.057\times 10^{-4}}{(0.453)^3}$$

$$= \tilde{C}_V^* + 4.3\times 10^{-4}R = 6.06 + 9\times 10^{-4}\text{cal/mol-K}.$$

Note that in this problem, $(\partial \tilde{C}_V/\partial p)_T$ amounts to 0.037 cal/mol-K-atm, but the pressure here is very low.

Example 9.3 Estimate the Joule–Kelvin coefficient for air at 80°F and 99 psia.

Solution: Treat air as if it were pure nitrogen at low pressure. The Joule–Kelvin coefficient can be expressed as

$$\mu = \left(\frac{\partial T}{\partial p}\right)_H = -\frac{(\partial T/\partial \tilde{H})_p}{(\partial p/\partial \tilde{H})_T} = \frac{RT^2}{\tilde{C}_p}\left[\frac{dB'}{dT} + p\frac{dC'}{dT} + \cdots\right].$$

In the limit of zero pressure, and with the corresponding-states Eq. 3.23 for the second virial coefficient, we have

$$\mu = \frac{\tilde{V}_c}{\tilde{C}_p^*}\left[-0.438 + \frac{1.762}{T/T_c} + \frac{2.271}{(T/T_c)^2}\right] = 0.7036\frac{\tilde{V}_c}{\tilde{C}_p^*},$$

since $T/T_c = 2.38$. With a heat capacity of $7R/2$, this yields

$$\mu = 0.7036\frac{2Z_cT_c}{7p_c} = \frac{0.7036\times 2\times 0.292\times 126}{7\times 33.5} = 0.221 \text{ K/atm},$$

in good agreement with the experimental value of 0.2192.

The Joule–Kelvin coefficient is zero for an ideal gas. In this respect, the ideal gas does not approximate the properties of real gases, even in the limit of zero pressure. In fact, the temperature at which μ is zero is different from the Boyle temperature where B' is zero (see Problem 9.2).

Problems

9.1 For the truncated virial equation
$$p\tilde{V}/RT = 1 + B'p,$$
show that

a. $p\left(\dfrac{\partial \tilde{V}}{\partial p}\right)_T = -\dfrac{RT}{p}.$

b. $\left(\dfrac{\partial \tilde{U}}{\partial p}\right)_T = -RT\dfrac{dB'T}{dT},$ independent of pressure.

c. $\tilde{C}_V = \tilde{C}_V^*(T) - Rp\dfrac{d^2(B'T^2)}{dT^2} - Rp^2\left(\dfrac{dB'T}{dT}\right)^2.$

9.2 The *inversion temperature* T_i is that for which the Joule–Kelvin coefficient $\mu = (\partial T/\partial p)_H$ is zero, in the limit of zero pressure. Thus, a gas above the inversion temperature will tend to heat up during a Joule–Kelvin (throttling) expansion.

    a. Show that the inversion temperature occurs at $T_i = 2a/Rb$ or $T_i/T_B = 2$ for a van der Waals fluid. Here, T_B is the Boyle temperature (see problem 3.4).

    b. Show that the inversion temperature occurs at $T_i = (5a/2Rb)^{2/3}$ or $T_i/T_B = 1.842$ for a Redlich–Kwong fluid.

    c. Show that the inversion temperature occurs at a reduced temperature of 5.05 or $T_i/T_B = 1.898$ for a fluid obeying the corresponding-states Eq. 3.23.

For nitrogen, $T_i/T_B = 1.905$, and for a fluid whose molecules follow the Lennard–Jones force law 3.16, the ratio T_i/T_B has the value 1.91.

Notation

a	van der Waals constant, atm-cm^6/mol^2, or Redlich–Kwong constant, K$^{1/2}$ -atm-cm^6 /mol^2
A	temperature-dependent parameter in Boyle's law, J/mol
b	van der Waals constant or Redlich–Kwong constant, cm^3/mol
B	second virial coefficient, cm^3/mol
B'	second virial coefficient in pressure expansion, atm^{-1}
C, D	third and fourth virial coefficients
C'	third virial coefficient in pressure expansion, atm^{-2}
$\tilde{C}_p$	molar heat capacity at constant pressure, J/mol-K
$\tilde{C}_V$	molar heat capacity at constant volume, J/mol-K
$\tilde{H}$	molar enthalpy, J/mol
p	pressure, N/m^2
R	universal gas constant, 8.3143 J/mol-K or 82.06 atm-cm^3/mol-K
T	thermodynamic temperature, K
T_B	Boyle temperature, K
T_i	inversion temperature, K
$\tilde{U}$	molar internal energy, J/mol
V	volume, m^3
$\tilde{V}$	molar volume, cm^3/mol

Z compressibility factor

μ Joule–Kelvin coefficient, K/atm

Subscript and special symbol

c critical point

*   ideal-gas state

Reference

1. M. W. Zemansky and H. C. Van Ness, *Basic Engineering Thermodynamics* (New York: McGraw-Hill Book Company, Inc., 1966).

Chapter 10

Surface Systems

Surfaces are of interest to chemical engineers for four principal reasons:

1. An important separation process involves adsorption on the extensive surface of porous carbon.

2. Surfaces are involved in all situations of interphase mass transfer. Surface tension is an important variable in determining how much interfacial area can be created in a sieve–plate column or a mixer, and the fluid flow near the interface can be strongly affected by variations in the surface tension due to composition and temperature variations.

3. Froth separation (see Chapter 22) has potential utility for recovering surfactants and other substances in the ocean by concentrating them in the region near the air–water surface.

4. In electrochemical and catalytic systems, chemical reactions occur at the surface.

Unfortunately, we can give only a superficial treatment of surfaces here. In this chapter, we shall characterize the surface by only two independent variables, the surface area and the temperature, and we shall try to ignore the substrate bulk phases. In Chapter 22, we return to multicomponent surface systems.

The Newman Lectures on Thermodynamics
John Newman and Vincent Battaglia
Copyright © 2019 Jenny Stanford Publishing Pte. Ltd.
ISBN 978-981-4774-26-0 (Hardcover), 978-1-315-10861-2 (eBook)
www.jennystanford.com

For our system, we can imagine a surface film stretched over a wire frame (see Fig. 10.1), with the end of the frame moveable so that we can change the area $\mathcal{A}$ of the surface. In this way, the system will be similar to a gas in a cylinder with a frictionless piston. Figure 10.2a is a plot of surface tension σ versus the area $\mathcal{A}$. This is to be analogous to the p–V plot in Fig. 10.2b for a PVT system.

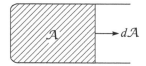

Figure 10.1 Surface film stretched over a wire frame, which allows the area of the surface to be varied. This constitutes a simple system.

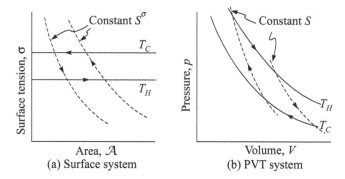

Figure 10.2 Surface tension–area plot and pressure–volume plot. Each plot shows two isotherms, with $T_H > T_C$, and two adiabatic curves. From these, we can construct a Carnot cycle.

For a system involving a single component, the surface tension will depend only on the temperature and will be independent of the area. Consequently, the isotherms are drawn as horizontal lines in Fig. 10.2a. Figure 10.3 shows that the surface tension σ between liquid and vapor water decreases with temperature and becomes zero at the critical point. This means that the lower-temperature isotherms are above the higher-temperature isotherms in Fig. 10.2a. This behavior is typical of the surface tension of a pure substance, and we shall represent σ as a function of temperature by the Eötvos equation[1]

$$\sigma = \sigma_0 \left(1 - \frac{T}{T_c} \right)^n ,$$ (10.1)

where σ_0 and n are constants, n being slightly larger than 1. For water, we shall use the values $n = 1.20$ and $\sigma_0 = 147$ dyne/cm.

The law of conservation of energy still reads

$$dU = đQ - đW.$$ (10.2)

For reversible processes, $đQ$ is equal to TdS, but the work now includes the work done to extend the surface against the surface tension as well as any pressure–volume work. Equation 10.2 becomes

$$dU = TdS - pdV + \sigma d\mathcal{A}.$$ (10.3)

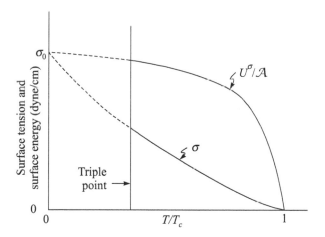

Figure 10.3 Surface tension σ and internal energy (per unit area) $U^\sigma/\mathcal{A}$ associated with the surface. The data are for water.

We now attempt, in our minds at least, to isolate the surface from the bulk phases adjacent to it, although this is not possible physically. First, we define the position of the surface. This is somewhere in the thin region in which the physical properties vary from one bulk

[1]In this equation, T_c is sometimes given a value slightly different from the critical temperature, in order to fit the surface tension better over a particular range of temperature. The resulting equation will then not reflect the fact that the surface tension must go to zero at the critical point.

phase to the other. Next we assign extensive properties, such as the volume and entropy, to one phase or the other. The entropy assigned to one bulk phase is its volume multiplied by its entropy per unit volume $\hat{S}/\hat{V}$. Finally, any of these extensive properties remaining is assigned to the surface. We shall denote the entropy thus assigned to the surface by S^σ, and similarly for the other extensive properties.

By this procedure, which is due to Gibbs, all the volume is assigned to the bulk phases, and as a result the volume V^σ of the surface is zero. We shall simplify matters by selecting the position of the *Gibbs surface* (the first of the above steps) in such a way that the mass of the surface is also zero.[2]

After subtracting the bulk phases, Eq. 10.3 becomes

$$dU^\sigma = TdS^\sigma + \sigma d\mathcal{A}, \tag{10.4}$$

the volume of the surface being zero. The surface Gibbs function is

$$G^\sigma = U^\sigma + pV^\sigma - TS^\sigma = U^\sigma - TS^\sigma, \tag{10.5}$$

and its variation is given by

$$dG^\sigma = -S^\sigma dT + \sigma d\mathcal{A}. \tag{10.6}$$

The Helmholtz free energy A^σ is identical to the Gibbs function G^σ for the surface, since the volume is zero. Similarly, H^σ equals U^σ.

Since σ depends only on T, Eq. 10.6 can be integrated immediately at constant temperature, with the result

$$G^\sigma = \sigma\mathcal{A}, \tag{10.7}$$

the integration constant being set equal to zero. We now see that the surface tension σ has the additional interpretation of being the surface Gibbs function per unit area $G^\sigma/\mathcal{A}$. The surface entropy can now be obtained with the aid of Eq. 10.6:

$$S^\sigma = -\left(\frac{\partial G^\sigma}{\partial T}\right) = -\mathcal{A}\frac{d\sigma}{dT}, \tag{10.8}$$

and the surface internal energy then is

---

[2]In multicomponent systems, one cannot, in general, select a surface position such that the mass of each component is zero in the surface. In those systems, one will need to be concerned also with surface concentrations Γ_i, expressed as moles of component *i* per unit area. The different possibilities, or *conventions*, for selecting the position of the Gibbs surface lead to much confusion in the treatment of multicomponent surface systems.

$$U^\sigma = G^\sigma + TS^\sigma = \mathcal{A}\left(\sigma - T\frac{d\sigma}{dT}\right). \qquad (10.9)$$

The Eötvos Eq. 10.1 allows us to obtain an explicit expression for the surface internal energy:

$$U^\sigma / \mathcal{A} = \sigma_0 \left[1 + (n-1)\frac{T}{T_c}\right]\left(1 - \frac{T}{T_c}\right)^{n-1}. \qquad (10.10)$$

Figure 10.3 also shows $U^\sigma/\mathcal{A}$ for water as a function of temperature. Since n is only slightly larger than 1, the surface internal energy varies slowly with temperature, except near the critical point, where it must go to zero. In this regard, $U^\sigma/\mathcal{A}$ or $H^\sigma/\mathcal{A}$ can be compared favorably with the enthalpy of vaporization (see Fig. 11.2).

Let us now turn to the thermal properties of the surface. The heat capacity at constant area is

$$C_{\mathcal{A}}^\sigma = T\left(\frac{\partial S^\sigma}{\partial T}\right)_{\mathcal{A}} = \left(\frac{\partial U^\sigma}{\partial T}\right)_{\mathcal{A}} = -\mathcal{A}T\frac{d^2\sigma}{dT^2}. \qquad (10.11)$$

One can see from Fig. 10.3 that $C_{\mathcal{A}}^\sigma$ is negative because U^σ decreases with increasing temperature. This is a surprising result, since we usually expect to encounter positive heat capacities. However, the heat capacity for a system that includes a significant amount of the bulk phase will be positive (see Problem 10.1).

Next we would like to determine how the temperature changes for a reversible, adiabatic extension of the surface. We interpret this to mean that S^σ is constant, that is, that the surface system is isolated from bulk phases. This temperature variation comes from the Maxwell relation corresponding to Eq. 10.4:

$$\left(\frac{\partial T}{\partial \mathcal{A}}\right)_{S^\sigma} = \left(\frac{\partial \sigma}{\partial S^\sigma}\right)_{\mathcal{A}} = \frac{(\partial\sigma/\partial T)_{\mathcal{A}}}{(\partial S^\sigma/\partial T)_{\mathcal{A}}} = -\frac{1}{\mathcal{A}}\frac{d\sigma/dT}{d^2\sigma/dT^2}. \qquad (10.12)$$

If the Eötvos Eq. 10.1 applies, this becomes

$$\left(\frac{\partial T}{\partial \mathcal{A}}\right)_{S^\sigma} = \frac{T_c - T}{\mathcal{A}(n-1)}, \qquad (10.13)$$

which is positive. Thus, stretching the surface would tend to raise the temperature, and one is now in a position to put curves of constant S^σ onto Fig. 10.2a.

Stretching the surface at constant temperature would involve absorption of heat by the surface since then

$$\left(\frac{đQ}{d\mathcal{A}}\right)_T = T\left(\frac{\partial S^\sigma}{\partial \mathcal{A}}\right)_T = -T\frac{d\sigma}{dT} \tag{10.14}$$

is positive. Consequently, a Carnot engine, absorbing heat at a high temperature and converting part of this to work, would trace its cycle in the direction shown by the arrows in Fig. 10.2.

In this chapter, simple surfaces provide an example of a thermodynamic system other than the PVT systems that we so commonly employ. They illustrate the fact that other systems can be taken through a Carnot cycle and used to establish the thermodynamic temperature scale. The surface systems are, in fact, simpler than the PVT systems because we have taken the surface tension to be independent of area. We do not have any rational basis for considering how the surface tension might depend on the area unless we are willing to treat the surface concentrations of species adsorbed on the surface (see the Gibbs adsorption equation 22.1).

An electrochemical cell might be another example of a thermodynamic system. If we disregard the volume, the pressure and volume can be replaced by the potential and the charge as independent variables. Without a detailed consideration of multicomponent systems, we have no rational basis for treating how the cell potential might depend on the charge. The entropy and the temperature remain as important variables, and this system can also be taken through a Carnot cycle. Note that the enthalpy and the Gibbs function do not retain their special significance for systems other than PVT systems.

Example 10.1 Droplets of liquid B are dispersed in liquid A. Find the pressure difference between these two phases.

Solution: Imagine cutting a droplet along the equator and make a force balance on one of the hemispheres (see Fig. 10.4). For a droplet of radius r_0, the surface force exerted on one hemisphere by the other is $2\pi r_0 \sigma$. The pressure force is the pressure difference $p_B - p_A$ multiplied by the *projected* area πr_0^2. Hence, the force balance becomes

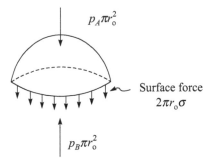

Figure 10.4 Force balance on the upper half of a droplet of fluid *B* in a medium of fluid *A*.

$$(p_B - p_A)\pi r_o^2 = 2\pi r_o \sigma$$

or

$$p_B - p_A = \frac{2\sigma}{r_o}.$$

Thus, the pressure within a droplet is greater than the pressure prevailing outside by an amount that increases as the drop becomes smaller.

This result is a special case of the *Laplace equation* (for surfaces)

$$p_B - p_A = \sigma\left(\frac{1}{r_1} + \frac{1}{r_2}\right),$$

where r_1 and r_2 are the *principal radii* of curvature of the surface. The more general equation can be applied locally to nonspherical surfaces such as a large mercury droplet resting on a glass plane, and it also applies to nonequilibrium situations, which might be found in the fluid mechanics of systems involving curved interfaces.

Example 10.2 Figure 10.5 shows a steady-flow process in which the interfacial area between two flowing phases is increased. Assume that the two phases are liquid and immiscible and that the surface tension is known as a function of temperature. In the absence of shaft work, how much pressure drop can be associated with a reversible, isothermal increase in the surface area, with the formation of droplets of radius r_o?

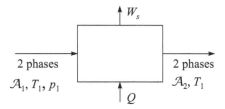

Figure 10.5 Steady-flow process for increasing the interfacial area in a liquid–liquid system.

Solution: For a reversible, isothermal process, the heat added is

$$Q = T_1 \Delta S,$$

it being assumed that a reversible heat source is available at the operating temperature T_1. The energy balance for a steady-flow process (Eq. 4.34) then reads

$$\Delta H = Q - W_s = Q = T_1 \Delta S$$

or

$$\Delta G = 0,$$

since the shaft work is zero. Since $G^\sigma = \mathcal{A}\sigma$, the energy balance becomes

$$m_A(\hat{G}_{A2} - \hat{G}_{A1}) + m_B(\hat{G}_{B2} - \hat{G}_{B1}) + (\mathcal{A}_2 - \mathcal{A}_1)\sigma = 0,$$

where B will denote the dispersed phase and A the continuous phase. For the bulk phases, Eq. 7.24 applies

$$\left(\frac{\partial \hat{G}}{\partial p}\right)_T = \hat{V},$$

and the densities are taken to be constant. This gives

$$m_A \hat{V}_A(p_{A2} - p_1) + m_B \hat{V}_B(p_{B2} - p_1) + (\mathcal{A}_2 - \mathcal{A}_1)\sigma = 0,$$

it being recognized that the pressure p_{B2} within the droplets is greater than the pressure p_{A2} in the continuous phase (see Example 10.1) by the amount

$$p_{B2} - p_{A2} = \frac{2\sigma}{r_0}.$$

We neglect the initial interfacial area $\mathcal{A}_1$. Substitution for p_{B2} in the energy balance and division by the volume yields

$$(1-\varepsilon)(p_{A2}-p_1)+\varepsilon\left(\frac{2\sigma}{r_0}+p_{A2}-p_1\right)+a\sigma=0,$$

where ε is the volume fraction of the dispersed phase and a is the specific interfacial area (that is, the area per unit volume). Rearrangement gives

$$p_1-p_{A2}=\sigma a+\frac{2\sigma\varepsilon}{r_0}=\frac{5\varepsilon\sigma}{r_0},$$

where we have lastly used the relationship

$$a=3\varepsilon/r_0$$

between the volume fraction and the specific interfacial area in a system dispersed into droplets of radius r_0.

Problems

10.1 The heat capacity of a surface between a pure liquid and its vapor can be expressed as

$$C_{\mathcal{A}}-C_{\mathcal{A}}^{\sigma}+\frac{dU_0}{dT}=-\mathcal{A}T\frac{d^2\sigma}{dT^2}+\frac{dU_0}{dT},$$

where U_0 represents the internal energy of the bulk liquid, which must be associated with the surface. For water at 0°C, determine whether $C_{\mathcal{A}}$ is positive or negative. Assume that the surface is 10^{-7} cm thick, and take the density and heat capacity of the bulk liquid to be 1 g/cm^3 and 4.18 J/g-K, respectively. For the surface, use the Eötvos equation with the parameters below Eq. 10.1.

10.2 Is the surface tension of a pure substance, like water, dependent on pressure?

10.3 In this chapter, a constant-pressure process is a constant-temperature process for a pure substance, since two phases are present. Is dQ equal to dH for such a process? (Compare Eq. 4.21 and Example 7.2.)

10.4 Draw isotherms and adiabatic curves on a potential–charge diagram for an electrochemical cell. The electrical work is $dW = -\Phi dq$, where Φ is the cell potential and q is the charge. Assume that the potential depends only on the temperature.

Discuss, with this diagram, how a Carnot power cycle would operate. Consider the case where the potential increases with temperature and do not violate the second law of thermodynamics. You may neglect volume changes for this system. Show that the slope of an adiabatic curve is

$$\left(\frac{\partial \Phi}{\partial q}\right)_S = \frac{T}{C_q}\left(\frac{d\Phi}{dT}\right)^2,$$

where C_q is the heat capacity at constant charge.

10.5 As an example of a gravitational system, take a lead weight that can be raised in a gravitational field and which can be heated from one temperature to another, having a heat capacity C. Volume changes are negligible, and the pressure and volume can be regarded to have been replaced by the weight w and the position z. The weight is independent of the temperature. Why cannot a gravitational system be taken through a Carnot cycle?

10.6 Air is forced through sieve plates, having holes of radius r_h, to form bubbles of radius r_o in a stagnant tank of water (see Fig. 10.6). On the basis of Example 10.1, we expect that a pressure drop of at least $2\sigma/r_h$ (between the water and the air below the plate) will be required. Derive an expression for the reversible pressure drop required to form bubbles of radius r_o and compare with the expected value of $2\sigma/r_h$. Assume isothermal operation and neglect any evaporation of water into the air. What is the minimum size of bubbles that can be created with holes of radius r_h (using the expected pressure drop of $2\sigma/r_h$)? Discuss the mechanism by which irreversibilities would account for the difference between the observed pressure and the reversible pressure when air is slowly forced through the plate.

10.7 Trace the development of a Carnot cycle for a system like that in Fig. 10.1, with inclusion of substrate fluid as in Problem 10.1. What modifications, if any, would be appropriate for Fig. 10.2a?

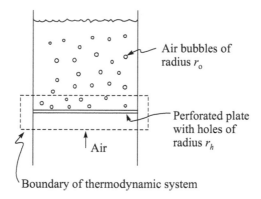

Figure 10.6   Sieve–plate system for producing small air bubbles.

Notation

a specific interfacial area, m^{-1}
A Helmholtz free energy, J
$\mathcal{A}$ surface area, m^2
$C_{\mathcal{A}}$ heat capacity at constant area, J/K
C_q heat capacity at constant charge, J/K
G Gibbs function, J
H enthalpy, J
m mass, kg
n constant
p pressure, N/m^2
q electric charge, C
Q heat, J
r_o radius of droplet, m
S entropy, J/K
T thermodynamic temperature, K
U internal energy, J
V volume, m^3
w weight, N
W work, J
W_s shaft work, J
z position, m
Γ_i surface concentration of species i, mol/m^2
σ surface tension, N/m

σ_0 constant, N/m

Φ cell potential, V

Subscripts, superscript, and special symbol

c critical point

C cold

H hot

σ surface

$\wedge$ per unit mass

Chapter 11

Phase Transition and Thermodynamic Diagrams

11.1 Phase Transition

The phase transition between a vapor and a liquid is depicted in Fig. 11.1. We are interested in studying such processes because

- This is a simple example of phase equilibrium. Multicomponent phase equilibria are of vital interest to chemical engineers in the analysis of separation operations.
- A phase transition complicates the procedure for the calculation of thermodynamic properties.
- Refrigeration processes (see Chapter 13) frequently involve cycling a substance through the phase transition from vapor to liquid and *vice versa*.

The vaporization process in Fig. 11.1 is carried out at constant temperature and pressure. Conservation of energy tells us that

$$\Delta U = Q - W. \tag{11.1}$$

For a reversible process, we can write

$$W = p\Delta V \tag{11.2}$$

The Newman Lectures on Thermodynamics
John Newman and Vincent Battaglia
Copyright © 2019 Jenny Stanford Publishing Pte. Ltd.
ISBN 978-981-4774-26-0 (Hardcover), 978-1-315-10861-2 (eBook)
www.jennystanford.com

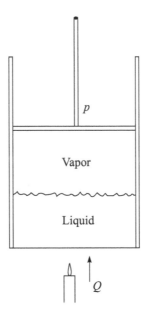

Figure 11.1 Vaporization of a liquid at constant pressure.

and

$$Q = T\Delta S, \qquad (11.3)$$

since both p and T are constants. Substitution into Eq. 11.1 gives

$$\Delta U = T\Delta S - p\Delta V \qquad (11.4)$$

or

$$\Delta(U - TS + pV) = \Delta G = 0. \qquad (11.5)$$

The change in the Gibbs free energy for the reversible vaporization is seen to be zero. Since

$$\Delta G = n_{\text{evap}}(\tilde{G}_{\text{vap}} - \tilde{G}_{\text{liq}}), \qquad (11.6)$$

where n_{evap} is the number of moles of the substance that are evaporated, it follows that

$$\tilde{G}_{\text{vap}} = \tilde{G}_{\text{liq}}. \qquad (11.7)$$

When two phases are in equilibrium, the Gibbs function (on either a molar or a mass basis) must be the same for the two phases.

Equation 11.7 is an example of the more general relation describing phase equilibrium in multicomponent systems. The chemical potential of a species i, denoted by μ_i, is identical to the partial molar Gibbs free energy, denoted by $\bar{G}_i$ and defined as

$$\bar{G}_i = \left(\frac{\partial G}{\partial n_i} \right)_{T,p,n_j \atop j \neq i}. \tag{11.8}$$

This is the partial derivative of the Gibbs function with respect to the number n_i of moles of species i, with temperature and pressure and the number of moles of any other species j being held constant during the differentiation.

The condition for equilibration of a species i between two phases α and β requires that

$$\mu_i^\alpha = \mu_i^\beta, \tag{11.9}$$

that is, the chemical potential of species i must be the same in the two phases. For a pure substance, $\mu_i = \tilde{G}$, and Eq. 11.9 reduces to Eq. 11.7. These concepts for multicomponent systems will be further developed in Part B.

Equation 11.7 permits us to treat the variation of the vapor pressure with temperature. For the vapor and the liquid, we can write

$$d\tilde{G}_{vap} = -\tilde{S}_{vap} dT + \tilde{V}_{vap} dp \tag{11.10}$$

and

$$d\tilde{G}_{liq} = -\tilde{S}_{liq} dT + \tilde{V}_{liq} dp. \tag{11.11}$$

In view of Eq. 11.7, we can write for the two-phase equilibrium

$$\left(\frac{dp}{dT} \right)_{sat} = \frac{\tilde{S}_{vap} - \tilde{S}_{liq}}{\tilde{V}_{vap} - \tilde{V}_{liq}} = \frac{1}{T} \frac{\tilde{H}_{vap} - \tilde{H}_{liq}}{\tilde{V}_{vap} - \tilde{V}_{liq}}, \tag{11.12}.$$

the last relationship following from the fact that $\Delta H = T\Delta S$ for the constant-pressure, constant-temperature process of evaporation (see Eq. 7.18).

Equation 11.12, known as the Clapeyron equation, is an important relationship between the vapor pressure and the enthalpy of vaporization. We can integrate Eq. 11.12 with the following three assumptions:

1. $\Delta \tilde{H}_{vap}$ is independent of temperature.
2. $\tilde{V}_{liq}$ is negligible compared to $\tilde{V}_{vap}$.
3. $\tilde{V}_{vap}$ can be expressed by the ideal-gas law, Eq. 3.8.

These three assumptions all become inaccurate near the critical point. However, they compensate largely for each other, as illustrated in Fig. 11.2 for water. With these assumptions, Eq. 11.12 becomes

$$\left(\frac{dp}{dT}\right)_{sat} = \frac{p}{T^2}\frac{\Delta\tilde{H}_{vap}}{R}, \tag{11.13}$$

and integration gives

$$\ln p_{sat} = \text{const} - \frac{\Delta\tilde{H}_{vap}}{RT}. \tag{11.14}$$

We prefer to write this in terms of critical properties as

$$\ln\frac{p_{sat}}{p_c} = \mathcal{A} - \mathcal{B}\frac{T_c}{T}, \tag{11.15}$$

since first of all, $\Delta\tilde{H}_{vap}/R$ is not truly a constant and, secondly, Eq. 11.15 represents a rather good corresponding-states correlation of the vapor pressure, as illustrated in Fig. 11.3. In order for a straight line to give the best representation over the entire range from the triple point to the critical point, $\mathcal{A}$ and $\mathcal{B}$ should be given the values

$$\mathcal{A} = 5.29 \text{ and } \mathcal{B} = 5.31 \tag{11.16}$$

(see Ref. [1], p. 170), although the actual curve must pass through the point $p = p_c$ at $T = T_c$.

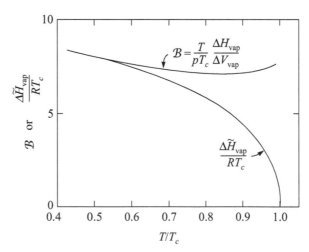

Figure 11.2 Graph indicating assumptions in the integration of the Clapeyron equation. The integration actually assumes that $T\Delta H_{vap}/p\Delta V_{vap}$ is a constant. The data are for water.

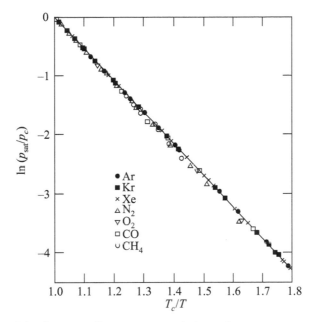

Figure 11.3 Corresponding-states correlation of vapor pressure with temperature. The slope of such a plot can be expressed as: slope $= -\mathcal{B} =$

$$\frac{d\ln(p_{sat}/p_c)}{d(T_c/T)} = \frac{-T^2}{T_c p_{sat}} \frac{dp_{sat}}{dT} = \frac{-T}{p_{sat}T_c} \frac{\Delta H_{vap}}{\Delta V_{vap}}$$

We have seen, particularly in Chapter 7, how to use the heat capacity and PVT data to calculate all the thermodynamic properties of a substance. In Chapter 9, we reiterated, with the aid of the virial equation, how this might be carried out for a gas or a vapor. For these purposes, the heat capacity was required, as a function of temperature, only in the ideal-gas limit.

We did not consider explicitly the possibility of phase transition, such as that sketched in Fig. 11.1. Can we integrate Eqs. 7.7, 7.10, 7.13, 7.33, and 7.35 through the phase transition in order to obtain the entropy, heat capacity, internal energy, and enthalpy for the liquid?

Consider the first Eq. 7.7:

$$\left(\frac{\partial S}{\partial V}\right)_T = \left(\frac{\partial p}{\partial T}\right)_V. \tag{11.17}$$

This equation is really the basis for using PVT data to calculate the pressure or volume dependence of all the thermal properties. In the two-phase region,

$$\left(\frac{\partial p}{\partial T}\right)_V = \left(\frac{dp}{dT}\right)_{sat}. \tag{11.18}$$

This can be visualized by studying Fig. 3.1, the pressure–volume diagram for water; a constant-volume process is a vertical line on that diagram. The right side of Eqs. 11.17 is thus a constant in the two-phase region, and integration at constant temperature yields

$$\tilde{S}_{vap} - \tilde{S}_{liq} = \left(\frac{dp}{dT}\right)_{sat} (\tilde{V}_{vap} - \tilde{V}_{liq}), \tag{11.19}$$

a result identical to Eq. 11.12. Thus, we see that it is a simple matter to integrate Eq. 7.7 or Eq. 7.13 through the phase transition and obtain the entropy, the internal energy, the enthalpy, and the Gibbs function for the liquid from the corresponding values for the vapor and the use of PVT data, in this case, the temperature dependence of the vapor pressure and the molar volumes of the saturated liquid and vapor.

One must, on the other hand, be cautious in trying to integrate Eq. 7.10 through the phase transition. The correct relationship between the heat capacities in the liquid and vapor phases at equilibrium should be

$$\frac{\tilde{C}_{V,liq} - \tilde{C}_{V,vap}}{T} = \left(\frac{d\tilde{V}_{vap}}{dT}\right)_{sat} \left[\left(\frac{\partial p}{\partial T}\right)_{V,vap} - \left(\frac{dp}{dT}\right)_{sat}\right]$$
$$- \left(\frac{d^2 p}{dT^2}\right)_{sat} (\tilde{V}_{vap} - \tilde{V}_{liq}) + \left(\frac{d\tilde{V}_{liq}}{dT}\right)_{sat} \left[\left(\frac{dp}{dT}\right)_{sat} - \left(\frac{\partial p}{\partial T}\right)_{V,liq}\right]. \tag{11.20}$$

This result is obtained by integrating Eq. 7.7 *through* the phase transition and then evaluating the heat capacity from its definition, Eq. 7.8.

Equations such as 7.35 would be difficult to integrate through the phase transition since the derivatives are self-contradictory; in the two-phase region, a constant pressure implies a constant temperature, and vice versa. Such an integration is unnecessary since the enthalpy of vaporization can be obtained from Eq. 11.12. Similarly, integration of Eq. 7.33 is unnecessary since the heat capacity at constant pressure for the liquid can be obtained from C_V (see Eq. 11.20) and the relationship between C_p and C_V (see Eq. 7.31).

11.2   Equilibrium

A state of equilibrium means that there is no tendency for spontaneous change within the system. However, it is implied in this statement that the system is isolated.

The condition for a spontaneous process is that the entropy change of the universe be positive:

$$\Delta S_{\text{universe}} > 0. \tag{11.21}$$

If there is no tendency for spontaneous change, then there must be no possible process that will increase the entropy; that is, the entropy of the universe has been maximized, subject to any constraints that might have been imposed.

The "universe" is used here in a restricted sense, to denote the system and its immediate surroundings that might interact with it.

For a system that is truly isolated, there are no surroundings and the condition 11.21 applies to the system itself. The system will change by itself until the entropy of the system has been maximized. Any possible process will increase the entropy, except a reversible process, which leaves the entropy of the system unchanged. An isolated vapor–liquid system is depicted in Fig. 11.4a. For any initial situation, the system will equilibrate in such a way that the entropy has been maximized. This isolated system is constrained to maintain constant values of the internal energy U and the volume V.

Frequently we like to contemplate systems that are constrained to equilibrate in contact with a reversible heat source at a temperature T and a reversible work source at a constant pressure p (see Fig. 11.4d). A spontaneous process here must obey condition 11.21, which can be expressed as

$$\Delta S + \Delta S_{\text{RHS}} + \Delta S_{\text{RWS}} > 0, \tag{11.22}$$

where no subscript denotes the system itself and the subscripts RHS and RWS denote the surroundings, the reversible heat source, and the reversible work source, respectively.

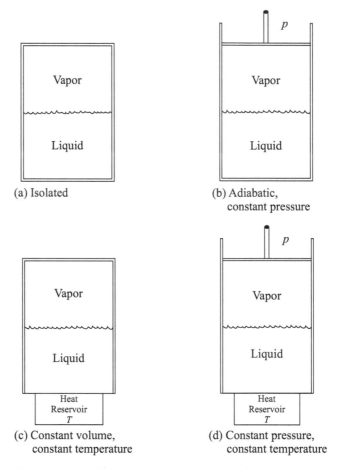

Figure 11.4   Systems with various constraints imposed.

The law of conservation of energy applies to any process, reversible or irreversible:

$$\Delta U = Q - W. \qquad (11.23)$$

The work against a constant pressure can be expressed as

$$W = p\Delta V, \qquad (11.24)$$

and the heat, being transferred from the reversible heat source, can be expressed as

$$Q = -T\Delta S_{RHS}. \qquad (11.25)$$

Furthermore, the entropy change for the reversible work source is zero. Substitution of these results into condition 11.22 gives

$$\Delta S - (\Delta U + p\Delta V)/T > 0. \tag{11.26}$$

Since p and T are constants, this can be rewritten as

$$\Delta(U + pV - TS) = \Delta G < 0. \tag{11.27}$$

On this basis, we expect that any spontaneous process in a system at constant pressure and temperature will result in a decrease in the Gibbs function for the *system*. We have translated the general criterion 11.21 for a spontaneous process into the restricted criterion 11.27. For these constraints of constant temperature and pressure, this latter criterion has the advantage that it refers to the systems rather than to the universe.

The system in Fig. 11.4d will equilibrate in such a way that the Gibbs function for the system is minimized. Any possible process will decrease the Gibbs function for the system, except a reversible process, which will leave the Gibbs function for the system unchanged. Ordinarily, we cannot carry out a reversible process at constant temperature and pressure. However, the vaporization in Fig. 11.4d or in Fig. 11.1 is such a process, and for this reversible process, we conclude that

$$\Delta G = 0, \tag{11.28}$$

in agreement with Eq. 11.5.

Conditions 11.27 or 11.28 for systems at constant temperature and pressure are convenient for establishing conditions for phase equilibrium or chemical equilibrium in reactive, multicomponent systems. The final equilibria attained for all the situations depicted in Fig. 11.4 are independent of the constraints imposed during the process, since at equilibrium there is no tendency for T, p, S, U, V, H, or G to change in any case. Thus, whenever a liquid is in equilibrium with its vapor,

$$\tilde{G}_{\text{vap}} = \tilde{G}_{\text{liq}}. \tag{11.29}$$

It should, perhaps, be emphasized that these last considerations and Eq. 11.28 apply only to PVT systems. In Example 7.2, the electrochemical cell could undergo a reversible process at constant temperature and pressure but for which ΔG was not equal to zero.

To illustrate the treatment of phase equilibrium, consider the van der Waals fluid, whose properties were developed in Example 7.4. Figure 11.5 shows the reduced pressure, Gibbs function, and Helmholtz free energy for a reduced temperature of 0.8. These isotherms, with a minimum and a maximum in the pressure and even a negative pressure, seem rather unlikely, and we wish to consider the possibility that the fluid might split into two phases, thereby producing a more stable situation.

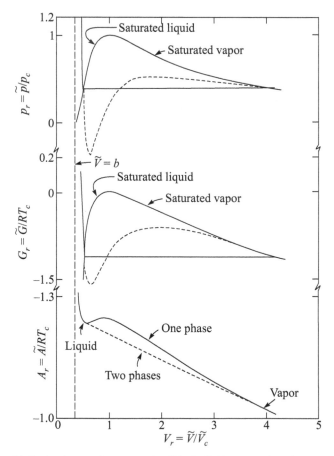

Figure 11.5 Isotherms for a van der Waals fluid at a reduced temperature of 0.8.

On the basis of the general criterion 11.21 of equilibrium, it can be shown that two coexistent phases must have the same pressure, the same temperature (see Example 11.2), and the same molar Gibbs function (see Eq. 11.29). Since Fig. 11.5 does not immediately reveal such a situation, we replot in Fig. 11.6 the reduced Gibbs function against the reduced pressure. The intersection on this graph identifies the conditions of saturated liquid and saturated vapor, and these loci are indicated in Fig. 11.5. For a mixture of liquid and vapor, the volume, Gibbs function, and Helmholtz free energy will each be a linear combination of the molar properties of the liquid and vapor. This will plot as a straight line in Fig. 11.5.

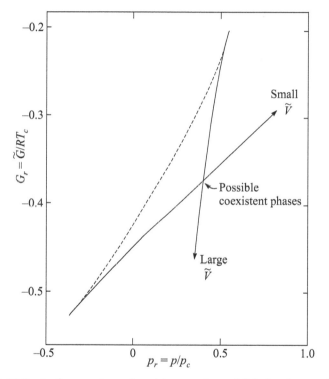

Figure 11.6 Isotherms at a reduced temperature of 0.8 showing coexistent phase with the same pressure and Gibbs function.

Which is more stable, one phase or two? One can use criterion 11.27 only at constant temperature and pressure. Indeed, at the saturation pressure, either Fig. 11.5 or Fig. 11.6 shows that G is lower for two phases than for one. Hence, a spontaneous process can result in a phase splitting.

More generally, we should like to use the constraints of constant volume and constant temperature (see Fig. 11.4c). (We like the constraint of constant temperature because we have already calculated the isotherms.) The curve for the Helmholtz free energy in Fig. 11.5 now shows isotherms for one phase and for two. For a fluid at constant volume (and temperature), the stable situation is that with the lower value of A, since a spontaneous process with these constraints must result in a decrease in A for the system (see Problem 11.1(b)).

Study of Fig. 11.5 now shows that phase separation requires that there be two points of inflection in the one-phase isotherm for the Helmholtz free energy. These points of inflection occur at minima and maxima in the isotherms for p and $\tilde{G}$ since

$$\left(\frac{\partial^2 A}{\partial V^2}\right)_T = -\left(\frac{\partial p}{\partial V}\right)_T = -\frac{1}{V}\left(\frac{\partial G}{\partial V}\right)_T. \qquad (11.30)$$

At the incipient point of phase separation, that is, at the critical point, the maximum and the minimum on the pressure isotherm come to coincide; this is the meaning of the criteria 3.9 for the critical point.

Proceeding in this manner at several reduced temperatures, we are able to produce Table 11.1 with the properties of the coexistent phases for condensation of a van der Waals fluid. Figure 11.7 is a plot of the vapor pressure. The van der Waals fluid is not a very good approximation to the properties of real fluids; the corresponding-states curve of the vapor pressure is carried over from Fig. 11.3. (We see that water also deviates considerably from the corresponding-states correlation, as could also be appreciated from the fact that the B values in Fig. 11.2 are not very close to the value 5.31 for the other substances.)

These calculations can be repeated for a fluid obeying the Redlich–Kwong equation of state (3.13). The properties of the saturated phases, again for $\tilde{C}_V^* = 5R/2$, are given in Table 11.2, and the vapor pressure curve is included in Fig. 11.7.

Table 11.1 Properties of saturated vapor and liquid for a van der Waals fluid with a heat capacity of $\tilde{C}_V = 5R/2$.

T_r	P_r	V_r	H_r	S_r
0.40	0.00517	0.3864	−3.6607	−4.8213
		203.6291	−0.3604	3.4294
0.45	0.01313	0.3960	−3.4638	−4.3603
		89.1467	−0.1986	2.8957
0.50	0.02779	0.4068	−3.2616	−3.9390
		45.9838	−0.0453	2.4936
0.55	0.05158	0.4188	−3.0529	−3.5483
		26.6099	0.0974	2.1796
0.60	0.08687	0.4326	−2.8364	−3.1814
		16.7285	−0.2277	1.9254
0.65	0.13584	0.4485	−2.6104	−2.8328
		11.1763	0.3437	1.7120
0.70	0.20046	0.4672	−2.3729	−2.4972
		7.8111	0.4432	1.5257
0.75	0.28246	0.4896	−2.1208	−2.1697
		5.6430	0.5234	1.3558
0.80	0.38336	0.5174	−1.8499	−1.8448
		4.1725	0.5802	1.1928
0.85	0.50449	0.5534	−1.5534	−1.5148
		3.1276	0.6070	1.0268
0.90	0.64700	0.6034	−1.2180	−1.1670
		2.3488	0.5909	0.8429
0.95	0.81188	0.6841	−0.8112	−0.7703
		1.7271	0.4994	0.6092
1.0	1.0	1.0	0.0	0.0

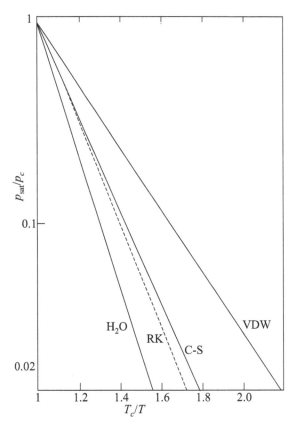

Figure 11.7   Vapor pressure curves for a van dar Waals fluids and a Redlich–Kwong fluid. Data for water and the fluids in Fig. 11.3 are also shown.

11.3   Thermodynamic Diagrams

To facilitate engineering calculations, tables and diagrams of thermodynamic properties are prepared for industrially important substances. These tables and diagrams contain information on five quantities, temperature T, pressure p, volume V, enthalpy H, and entropy S. From these quantities, one can easily calculate the internal energy U, the Helmholtz free energy A, and the Gibbs function G. We might notice that such tables and diagrams are not the most *accurate* way to record thermodynamic data. More accurate would be tables of the isothermal compressibility k, the coefficient of thermal expansion β, and the heat capacity $\hat{C}_p$. To obtain these quantities by differentiation of the tables and diagrams would be less accurate than the calculation of V, H, and S by integration of the differential quantities.

Table 11.2 Properties of saturated vapor and liquid for a Redlich–Kwong fluid with a heat capacity of $\tilde{C}_V^* = 5R/2$.

T_r	P_r	V_r	H_r	S_r
0.45	0.00046	0.2993	−6.8947	−9.4678
		2922.5246	0.4502	6.8541
0.50	0.00226	0.3082	−6.2748	−8.1608
		661.1035	0.6201	5.6289
0.55	0.00771	0.3185	−5.7023	−7.0700
		210.9838	0.7820	4.7195
0.60	0.02047	0.3306	−5.1636	−6.1347
		85.3664	0.9302	4.0217
0.65	0.04526	0.3450	−4.6477	−5.3130
		40.7953	1.0589	3.4663
0.70	0.08742	0.3625	−4.1446	−4.5747
		21.9403	1.1619	3.0061
0.75	0.15249	0.3842	−3.6448	−3.8962
		12.8349	1.2328	2.6074
0.80	0.24589	0.4119	−3.1375	−3.2574
		7.9622	1.2636	2.2439
0.85	0.37273	0.4491	−2.6081	−2.6377
		5.1289	1.2422	1.8920
0.90	0.53780	0.5031	−2.0323	−2.0097
		3.3567	1.1467	1.5225
0.95	0.74548	0.5949	−1.3530	−1.3167
		2.1542	0.9220	1.0781
1.0	1.0	1.0	0.0	0.0

Five different ways of plotting the thermodynamic data are popular:

1. *p–V* diagram. Figure 3.1 is such a diagram for water, although lines of constant S and H would be required for completeness.

This diagram is appealing because of the direct mechanical significance of pressure and volume. The reversible work, being the integral of pdV, is the area under a curve for the process on this diagram (although a logarithmic plot such as Fig. 11.8 would be necessary to cover adequately the relevant ranges of p and V), and the area within a closed curve is the work for a cyclic process. For a reversible, adiabatic process (constant entropy), the enthalpy change (equal to $-W_s$) is the integral of Vdp. This is, again, an area on the p–V diagram.

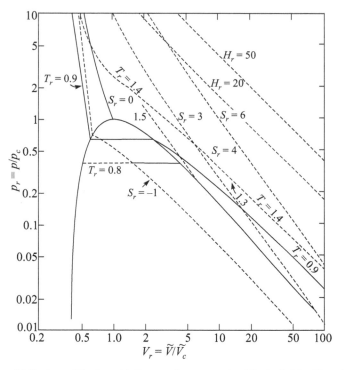

Figure 11.8 Logarithmic p–V diagram for a van der Waals fluid with a heat capacity of $\tilde{C}_V = 5R/2$.

2. p–T diagram. Figure 3.2 is such a diagram for water, although it would be more useful if a logarithmic scale were used for the pressure. Since neither pressure nor temperature is extensive, this is the only diagram that shows a two-phase curve rather than a two-phase region.

3. *T–S* diagram. A Carnot cycle is a rectangle on this diagram; see Fig. 11.9. The *T–S* diagram is the thermal analogue of the *p–V* diagram since the reversible heat is the integral of *TdS*. Thus, the area within a closed curve is the heat for a reversible, cyclic process, and this is equal to the area on the *p–V* diagram since then $W = Q$.

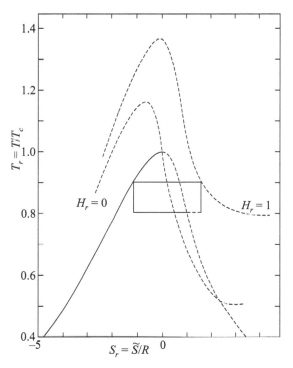

Figure 11.9 *T–S* diagram for a van der Waals fluid with $\tilde{C}_V = 5R/2$. The two Carnot cycles, which are rectangles in this diagram, are transcribed onto Figs. 11.8, 11.10, and 11.11.

4. *p–H* diagram, usually with a logarithmic scale for the pressure (see Fig. 11.10). The enthalpy is useful because for a steady flow process

$$\Delta H = Q - W_s.$$

Thus, for an adiabatic compressor or turbine, the enthalpy change is directly related to the shaft work; and, for a simple heat exchanger, the enthalpy change is directly related to the heat transferred. The *p–H* diagram is especially well suited for refrigeration work (see Chapter 13) since the evaporators and condensers operate at constant pressure and the Joule–Kelvin expansion is a constant-enthalpy process.

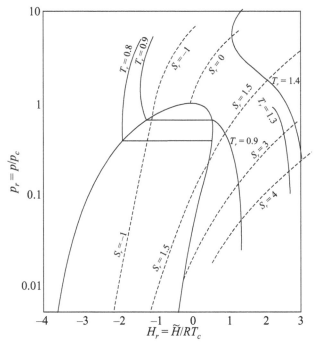

Figure 11.10 Semi-logarithmic *p–H* diagram for a van der Waals fluid with a heat capacity of $\tilde{C}_V = 5R/2$.

5. *H–S* or *Mollier* diagram (see Fig. 11.11). Here, the enthalpy has the advantage cited above, and the turbine in an efficient power cycle will involve a nearly-constant-entropy process.

The Appendix contains a *p–H* diagram for chlorodifluoromethane (Freon 22 refrigerant), compliments of Chemours. *H–S* and *p–H* diagrams of water and carbon dioxide can be found on the internet, for example, this link https://www.ohio.edu/mechanical/thermo/property_tables/index.html.

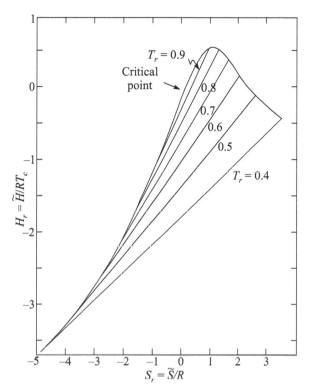

Figure 11.11 *H–S* diagram for a van der Waals fluid with a heat capacity of $\tilde{C}_V = 5R/2$.

11.4 Sources of Data

It is always a challenge to put one's hands quickly on just the right data for a particular chemical. A detailed library search may yield several sets of original data, which do not quite agree. For common substances, the results have been analyzed carefully, smoothed, and presented in tabular or graphical form. For less common substances or for quick answers, do not hesitate to use corresponding-states correlations. The modern student will quickly learn to find the available data on the Internet.

Example 11.1 Estimate the heat capacity of saturated liquid water at 20°C, using Eq. 11.20 as a basis.

Solution: Write

$$\left(\frac{d\tilde{V}_{\text{liq}}}{dT}\right)_{\text{sat}} = \left(\frac{\partial \tilde{V}}{\partial T}\right)_p + \left(\frac{\partial \tilde{V}}{\partial p}\right)_T\left(\frac{dp}{dT}\right)_{\text{sat}} = \tilde{V}_{\text{liq}}\left[\beta - k\left(\frac{dp}{dT}\right)_{\text{sat}}\right]_{\text{liq}}$$

and

$$\left(\frac{\partial p}{\partial T}\right)_V = -\frac{\beta}{k}$$

(see Eq. 1.5). Similar equations apply to the vapor phase. Now

$$\left(\frac{dp}{dT}\right)_{\text{sat}} = \frac{\mathcal{B}T_c p_{\text{sat}}}{T^2}$$

and

$$\frac{d^2 p_{\text{sat}}}{dT^2} = \frac{T_c p_{\text{sat}}}{T^3}\left[\frac{T}{T_c}\frac{d\mathcal{B}}{d(T/T_c)} + \mathcal{B}\left(\frac{\mathcal{B}T_c}{T} - 2\right)\right],$$

$\mathcal{B}$ being one of the quantities plotted for water in Fig. 11.2. Equation 11.20 now becomes

$$\tilde{C}_{V,\text{liq}} - \tilde{C}_{V,\text{vap}} = -\frac{p_{\text{sat}}}{T}\left(\tilde{V}_{\text{vap}} - \tilde{V}_{\text{liq}}\right)\frac{d\mathcal{B}}{d(T/T_c)}$$

$$-\left(\frac{\beta^2 T\tilde{V}}{k}\right)_{\text{liq}}\left[1 - kp\left(\frac{\mathcal{B}T_c}{\beta T^2}\right)^2\left(1 - kp - 2\frac{(1-\beta T)T}{\mathcal{B}T_c}\right)\right]_{\text{liq}}$$

$$+ \tilde{V}_{\text{vap}}\left[\frac{\beta^2 T}{k} + 2(1-\beta T)\frac{T_c p_{\text{sat}}\mathcal{B}}{T^2} + \frac{\mathcal{B}^2 p_{\text{sat}} T_c^2}{T^3}(kp_{\text{sat}} - 1)\right]_{\text{vap}}.$$

From the data in Fig. 11.2, $\mathcal{B} = 8.218$ and

$$\frac{d\mathcal{B}}{d(T/T_c)} = -4.903 \, .$$

With the liquid properties $\rho = 0.99823$ g/cm^3, $\beta = 2.07 \times 10^{-4}$ K^{-1}, and $k = 4.58 \times 10^{-11}$ cm^2/dyne, the contribution to the change in heat capacity from the second line in that equation amounts to

$$-0.118[1 - 0.096(1.0 - 0.104)] = -0.108 \text{ cal/mol-K}.$$

Under these conditions, the reduced temperature and reduced pressure are

$$T/T_c = 293.15/647.25 = 0.4529$$

and

$$p/p_c = 0.3388/3206.2 = 1.057 \times 10^{-4}.$$

At this low reduced pressure, the vapor is nearly ideal, and it is helpful to introduce the compressibility factor $Z = p\tilde{V}/RT$ for the vapor phase. The last line in the equation for $\Delta\tilde{C}_V$ then becomes

$$ZR\frac{\left[1+\dfrac{T}{Z}\left(\dfrac{\partial Z}{\partial T}\right)_p\right]^2}{1-\dfrac{p}{Z}\left(\dfrac{\partial Z}{\partial p}\right)_T} - 2BRT_c\left(\frac{\partial Z}{\partial T}\right)_p - Rp\left(\frac{BT_c}{T}\right)^2\left(\frac{\partial Z}{\partial p}\right)_T$$

$$\approx R\left[Z^2 + 2\left(Z - \frac{BT_c}{T}\right)pT\frac{dB'}{dT} - \left(\frac{BT_c}{T}\right)^2 B'p\right],$$

where we have used the truncated virial equation 3.25. From the corresponding-states correlation 3.23 of the second virial coefficient, we estimate that

$$B'p = -2.79 \times 10^{-4} \quad \text{and} \quad pT\frac{dB'}{dT} = 7.8 \times 10^{-4}.$$

The contribution of the vapor term is thus estimated to be

$$1.0646R = 2.114 \text{ cal/mol-K}.$$

The $\Delta\tilde{C}_V$ equation now is

$$\tilde{C}_{V,\text{liq}} - \tilde{C}_{V,\text{vap}} = -ZR\left(1 - \frac{\tilde{V}_{\text{liq}}}{\tilde{V}_{\text{vap}}}\right)\frac{d'B}{d(T/T_c)} - 0.108 + 2.114$$

$$= 9.733 + 2.007 = 11.740 \text{ cal/mol-K}.$$

Example 9.2 yielded the result $\tilde{C}_{V,\text{vap}} = 6.0609$ cal/mol-k . Hence, the heat capacity of the liquid is

$$\tilde{C}_{V,\text{liq}} = 17.800 \frac{\text{cal}}{\text{mol-k}} \quad \text{or} \quad \hat{C}_{V,\text{liq}} = 4.1369 \frac{\text{J}}{\text{g-k}}.$$

The difference between the heat capacities at constant pressure and at constant volume was found to be 0.0275 J/g-K in Example 7.1. Hence, our final estimate of the heat capacity of liquid water at 20°C is

$$\hat{C}_{p,\text{liq}} = 4.1644 \text{ J/g-k,}$$

which differs by 0.42% from the measured value of 4.1819 J/g-K.

The contributions to $\hat{C}_{p,\text{liq}}$ can be summarized as follows:

0.33639 $\hat{C}_V^*$ in the limit of zero pressure
0.00005   correction to p_{sat} (Example 9.2)
0.65166   correction from vapor to liquid (Example 11.1)
0.00657   conversion, const V to const p (Example 7.1)
——————
0.99467 $\hat{C}_{p,\text{liq}}$ in cal/g-K.

The readers should appreciate that the determination of liquid properties from ideal-gas properties by integration of PVT properties is frequently more difficult and less accurate than a direct measurement. It is also interesting to observe that the major contribution to the correction from vapor to liquid is the term involving the derivative of $\mathcal{B}$, a term that would be zero if the corresponding-states correlation 11.15 were strictly applicable with constant values of $\mathcal{A}$ and $\mathcal{B}$.

Example 11.2 Two identical bodies, each having the equation of state

$$U = CT,$$

are isolated as in Fig. 11.4a. (These can be regarded to have identical heat capacities and no essential pressure–volume characteristics.) If the initial temperature of body 1 is T_1° and that of body 2 is T_2° (where T_1° is greater than T_2°), show that the final temperatures of these two bodies, after equilibrium is reached, are equal.

Solution: The entropy change for any possible process is

$$\Delta S = \Delta S_1 + \Delta S_2 = C \ln\frac{T_1}{T_1^\circ} + C \ln\frac{T_2}{T_2^\circ}.$$

To attain equilibrium, this entropy change is to be maximized subject to the constraint that the internal energy of the isolated system is constant:

$$\Delta U = 0 = \Delta U_1 + \Delta U_2 = C\left(T_1 - T_1^{\circ}\right) + C\left(T_2 - T_2^{\circ}\right).$$

Elimination of T_1 from the expression for ΔS yields

$$\Delta S = C \ln \frac{T_1 T_2}{T_1^{\circ} T_2^{\circ}} = C \ln \frac{T_2 (T_1^{\circ} + T_2^{\circ} - T_2)}{T_1^{\circ} T_2^{\circ}},$$

which is plotted in Fig. 11.12. The maximum occurs when

$$\frac{\partial \Delta S}{\partial T_2} = 0 = \frac{C}{T_2} - \frac{C}{T_1^{\circ} + T_2^{\circ} - T_2}$$

or

$$T_2 = \frac{T_1^{\circ} + T_2^{\circ}}{2}.$$

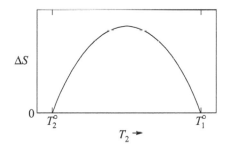

Figure 11.12 Entropy change as a function of the temperature of body 2, with a constraint of constant internal energy.

Substitution into the constraint of constant internal energy shows that the final equilibrium temperature of body 1 is also equal to this average of the two initial temperatures.

Example 11.3 Repeat Example 5.1, but use the actual properties of steam in the estimation of the flow rate through the adiabatic compressor.

Solution: From the steam tables, one finds that the enthalpy change for the steam is

$$\Delta \hat{H} = 1329.6 - 1236.9 = 92.7 \text{ Btu/lb} = 51.5 \text{ cal/g}.$$

The flow rate is, therefore, given by

$$\frac{1000 \text{ J}}{\text{s}} \frac{\text{cal}}{4.184 \text{ J}} \frac{\text{g}}{51.5 \text{ cal}} = 4.641 \text{ g/s}.$$

The answer obtained in Example 5.1, by treating steam as an ideal gas, differs from this result by only 0.3%.

Example 11.4 In the situation of Example 11.3 above, calculate the ratio of the actual work to the work that would have been required for a reversible adiabatic compression to the same final pressure.

Solution: For a reversible adiabatic compression, $\Delta \hat{S} = 0$. Therefore, interpolate in the steam tables at 75 psia to find the condition such that $\hat{S}_{2,rev} = \hat{S}_1 = 1.7758$ Btu/lb-deg R. We obtain then $T_{2,rev} = 571.55°F$ and $\hat{H}_{2,rev} = 1315.7$ Btu/lb. Consequently, the work ratio required is

$$\frac{W_s}{W_{s,rev}} = \frac{\Delta H}{\Delta H_{rev}} = \frac{92.7}{1315.7 - 1236.9} = 1.177.$$

This result is identical to that of Example 5.2, although the calculated temperature $T_{2,rev}$ is slightly higher (the value 569.91°F having been obtained when the steam was treated as an ideal gas).

Problems

11.1 (a) For Fig. 11.4b, subject to the constraints of constant enthalpy and pressure, show that $\Delta S > 0$ for any spontaneous process.

(b) For Fig. 11.4c, subject to the constraints of constant volume and temperature, show that $\Delta A < 0$ for any spontaneous process.

11.2 (a) Show that in an *H–S* diagram, an isobar has a slope equal to *T*.

(b) Show that in an *H–S* diagram, an isotherm has a slope equal to $T - 1/\beta$, where β is the coefficient of thermal expansion.

(c) In a *p–H* diagram, show that the curve for a reversible, adiabatic process (a curve of constant *S*) has a slope of $1/V$ and that such a curve has a slope of $0.4343/p\hat{V}$ on a plot of log *p* versus $\hat{H}$. These curves, therefore, have a continuous slope at the phase boundary in Fig. 11.10.

Notation

A	Helmholtz free energy, J
$\mathcal{A}$	constant
B'	second virial coefficient in pressure expansion, atm^{-1}
$\mathcal{B}$	constant or $(T/p_{sat}T_c)\Delta H_{vap}/\Delta V_{vap}$
C	heat capacity, J/K
$\hat{C}_p$	heat capacity at constant pressure, J/g-K
G	Gibbs function, J
H	enthalpy, J
k	isothermal compressibility, m^2/N
n_i	number of moles of species i, mol
p	pressure, N/m^2
Q	heat, J
R	universal gas constant, 8.3143 J/mol-K or 82.06 atm-cm^3/mol-K
S	entropy, J/K
T	thermodynamic temperature, K
U	internal energy, J
V	volume, m^3
W	work, J
Z	compressibility factor
β	coefficient of thermal expansion, K^{-1}
μ_i	chemical potential, J/mol
ρ	density, g/cm^3

Subscripts, superscripts, and special symbols

c	critical point
liq	liquid
r	reduced property
sat	saturation condition
vap	vapor or vaporization
α, β	phases α and β
$\sim$	per mole
$\wedge$	per unit mass
$-$	partial molar

References

1. E. A. Guggenheim, *Thermodynamics: An Advanced Treatment for Chemists and Physicists* (Amsterdam: North-Holland Pub. Co., 1959).

2. Joseph H. Keenan and Joseph Kaye, *Gas Tables* (New York: John Wiley & Sons, Inc., 1948).

3. J. B. Maxwell, *Data Book on Hydrocarbons* (New York: D. Van Nostrand Company, Inc., 1950).

4. Joseph H. Keenan and Frederick G. Keyes, *Thermodynamic Properties of Steam* (New York: John Wiley & Sons, Inc., 1936).

5. R. W. Bain, *Steam Tables 1964* (Edinburgh: Her Majesty's Stationery Office, 1964).

6. J. H. Dymond and E. B. Smith, *The Virial Coefficients of Gases* (Oxford: Clarendon Press, 1969).

7. F. Din (ed.), *Thermodynamic Functions of Gases. Volume 1, Ammonia, Carbon Dioxide and Carbon Monoxide* (London: Butterworths Scientific Publications, 1956). *Volume 2, Air, Acetylene, Ethylene, Propane, and Argon* (London: Butterworths Scientific Publications, 1956). *Volume 3, Methane, Nitrogen, Ethane* (London: Butterworths Scientific Publications, 1961).

8. Frederick D. Rossini, Kenneth S. Pitzer, Raymond L. Arnett, Rita M. Braun, and George C. Pimentel, *Selected Values of Physical and Thermodynamic Properties of Hydrocarbons and Related Compounds* (American Petroleum Institute Research Project 44, Pittsburgh: Carnegie Press, 1953).

9. Robert H. Perry and Cecil H. Chilton (eds.), *Chemical Engineers' Handbook* (New York: McGraw-Hill Book Company, 1973), pages 3–150 to 3–207.

10. Daniel R. Stull, "Vapor pressure of pure substances—Organic compounds," *Industrial and Engineering Chemistry*, **39**, 517–550, (1947).

Chapter 12

Work Processes and Cycles

One question that thermodynamics has classically treated is the determination of the maximum work that can be obtained (or the minimum work required) when a process is carried out subject to one or more constraints. For example, the Carnot cycle tells how much work can be obtained when heat is extracted from a hot reservoir and waste heat is rejected to a cold reservoir. Here, it is important to be convinced that the maximum work is always obtained when the process is reversible. Example 12.1 and Problems 12.1a and 12.2 illustrate this for several different constraints.

For a reversible process, the conservation of energy can be expressed as

$$dU = đQ - đW = TdS - đW \qquad (12.1)$$

or

$$dA = dU - TdS - SdT = -SdT - đW, \qquad (12.2)$$

where A is the Helmholtz free energy (see Eqs. 7.14 and 7.17). Thus, for any *isothermal*, reversible process,

$$đW = -dA \qquad (12.3)$$

or

$$W = -\Delta A, \qquad (12.4)$$

The Newman Lectures on Thermodynamics
John Newman and Vincent Battaglia
Copyright © 2019 Jenny Stanford Publishing Pte. Ltd.
ISBN 978-981-4774-26-0 (Hardcover), 978-1-315-10861-2 (eBook)
www.jennystanford.com

and the Helmholtz free energy can be associated with the maximum work that can be extracted from the system during an isothermal process.

At this point, we should probably recall the expression 4.34 for conservation of energy for a steady-flow process. For a reversible, isothermal process, the heat can be written as $T\Delta S$, and we find that the shaft work is $W_s = T\Delta S - \Delta H = -\Delta G$. For this reason, the Gibbs function can be associated with the shaft work for isothermal, reversible steady-flow processes. (This result finds application in Examples 10.2 and 17.2.) A similar association can be made for isothermal, constant-pressure processes, where Eq. 4.22 would then become $W' = -\Delta G$. The difference between the Helmholtz free energy and the Gibbs function in this context is similar to the difference between the internal energy and the enthalpy. A certain amount of pressure–volume work—the flow work or the work done against the constant pressure of the atmosphere—is accounted for separately with G and H, so that W_s or W' represents work available for other external processes. The Gibbs function thus finds more frequent application and is more frequently tabulated than the Helmholtz free energy.

12.1   Steam Power Plant

The only power cycle worthy of our attention here is that used in a typical steam plant to produce electric power. There are also the air-standard Otto cycle and the air-standard diesel cycle, used by mechanical engineers to simulate or approximate in some sense the internal combustion engine. The Carnot cycle does not find practical application in the production of power from heat.

Figures 12.1 and 12.2 illustrate a steam power cycle on T–S and p–H diagrams for water, and Fig. 12.3 is a schematic diagram of the process. Data for the condition of the steam at various points in the cycle are as follows:

S. No.	Location	T	p	$\hat{V}$	$\hat{H}$	$\hat{S}$
		°F	psia	ft³/lb	Btu/lb	Btu/lb-°R
1.	After condenser	79.58	0.50	0.01608	47.60	0.0924
2.	After pump	79.95	2000	0.01598	53.55	0.0924

S. No.	Location	T °F	p psia	$\hat{V}$ ft³/lb	$\hat{H}$ Btu/lb	$\hat{S}$ Btu/lb-°R
3.	After preheater	635.82	2000	0.0257	671.7	0.8619
4.	After boiler	635.82	2000	0.1878	1135.1	1.2849
5.	After superheater	1026.1	2000	0.4035	1490.7	1.5713
6.	After first turbine	312.03	80	5.472	1183.1	1.6207
7.	After reheater	1063.1	80	11.304	1564.4	1.9654
8.	After second turbine	79.58	0.50	641.4	1096.4	2.0372

Exhaust steam from the second turbine is condensed at an absolute pressure of 0.5 psi. This low pressure is used in order to reduce as much as possible the temperature at which the heat Q_C is rejected. This means that the condenser is operated below atmospheric pressure, and, consequently, it is important to have a means for removing noncondensable components such as air, which might leak into the system.

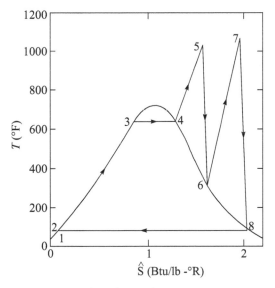

Figure 12.1 Steam power plant depicted in a temperature–entropy diagram for water. This might be called a Rankine cycle with superheater and reheater. A true Rankine cycle would eliminate the superheater and reheater. The turbine would then operate in the two-phase region and might be represented as a vertical line descending from point 4.

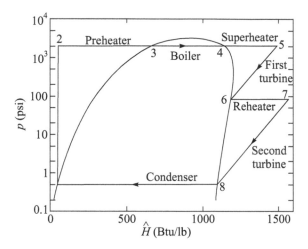

Figure 12.2 Steam power plant depicted in a pressure–enthalpy diagram for water.

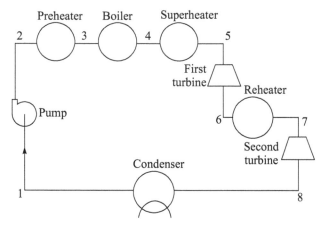

Figure 12.3 Steam power plant with two turbines. These turbines and the pump are supposed to operate adiabatically. Heat from burning the fuel is added to the preheater, boiler, superheater, and reheater, while waste heat is rejected to cooling water in the condenser.

The condensate is pumped into the preheater at the high pressure of 2000 psia. The points 1 and 2 are hardly distinguishable on the T–S diagram, in contrast to the p–H diagram. This means that the isobars for the liquid are grouped very close together on the T–S diagram and nearly coincide with the curve for the saturated liquid.

A preheater may not be used, in which case the liquid from the pump enters the boiler directly. This is necessarily an irreversible process since cold liquid would be mixed with hot liquid. However, the heating will be irreversible anyway if only one flame temperature is available. Efficiency can be gained by cooling the flue gases in the preheater before venting them up the stack.

The steam from the boiler is superheated at the same pressure, and the system is designed carefully so that the steam will *not* condense in the turbine. Liquid in the turbine causes erosion of the blades and catastrophic failure. (Some special turbines are designed to operate with wet steam, but they are not common.) With superheat, the highest temperature is considerably above the evaporation temperature. Material considerations limit the maximum temperature to about 1150°F, and most power plants do not reach the temperature of 1063°F used in this example.

The turbines in Fig. 12.3 were taken to operate adiabatically with an efficiency of 80%, that is, they produce 80% as much power as a reversible, adiabatic expansion over the same pressure range. Consequently, the lines 5–6 and 7–8 in Fig. 12.1 do not drop vertically or isentropically. Instead, the entropy of the steam is shown to increase in each turbine.

The steam in the first turbine is shown to reach the saturated state at a pressure of 80 psia. Since the temperature is still 312°F but condensate cannot be permitted in the turbine, the steam is reheated to a high temperature and passed through a second turbine whose exhaust is at a sufficiently low temperature. In Problem 12.3 you are asked to calculate the efficiency if the second turbine is eliminated and the steam at point 6 is condensed.

For the adiabatic turbines, as well as the adiabatic pump, the shaft work is calculated from Eq. 4.34 with $Q = 0$:

$$\Delta H = -W_s. \tag{12.5}$$

The required values of $\widehat{H}$ are readily given by Fig. 12.2. The first turbine produces 307.6 Btu of work per pound of flowing steam and the second yields 468, while the pump requires shaft work of only 5.9 Btu/lb. The net work is thus 769.7 Btu/lb. Notice the large difference in volume between the vapor from the second turbine and the liquid being pumped, a factor of over 40,000. Relatively small amounts of work are involved in handling a liquid, since its volume is small.

The heat Q_H extracted from the hot source is the sum of the values for the preheater, boiler, superheater, and reheater. Since no shaft work is involved in these units, Eq. 4.34 relates the heat to the enthalpy change:

$$\Delta H = Q. \tag{12.6}$$

Adding 1437.2 for the first three components and 381.3 for the reheater, we obtain a total of $\hat{Q}_H = 1818.5$ Btu/lb. The efficiency of this power cycle is thus calculated to be

$$\eta = \frac{W}{Q_H} = \frac{769.7}{1818.5} = 42.3\%. \tag{12.7}$$

This does not account for the thermal losses in the heaters, which might amount to 15% of the heating value of the fuel going up the stack or being lost through the furnace walls.

The conditions chosen for this power plant involve relatively high temperatures and pressures in the heaters and a relatively low temperature and pressure in the condenser. The efficiency of 42.3% is, therefore, toward the high side of the range that is likely to be encountered. However, it is not unreasonable for modern power plants.

12.2 Compression of Gases

Chemical reactions generally occur more rapidly at higher temperatures, and the reactants are frequently gases. But an unfavorable equilibrium conversion may dictate that the reactor be operated at high pressure. In such a case, the energy and cost to compress the reactants can become appreciable. And the gas must be expected to become hot during the compression process.

Even the simpler problem of transmission of natural gas over distances of hundreds of kilometers requires consideration of the optimum pressure and pipe diameter as well as the energy to overcome friction. In a system for solar heating of a home, one can choose among air and water and expensive liquids for the heat-transfer medium. The low heat capacity and high cost of blowing air must be weighed against the freezing, leaking, and corrosion problems of water and the high cost of liquids besides water.

Refrigeration involves the use of mechanical energy to transfer heat from a cold region to a hot region. The mechanical energy generally goes directly into the compression of a gas.[1] (See Chapter 13.)

Gas liquefaction, involving compression as an important process step, can be used to separate air into its components since oxygen boils at 90.2 K and nitrogen boils at 77.3 K. Argon and other noble gases are also recovered as by-products, while carbon dioxide is discarded. Liquid nitrogen is a convenient source of a moderately low temperature for processes such as the rapid freezing of food. To perform experiments near absolute zero, it is essential to liquefy hydrogen (boiling at 20.4 K) and helium (boiling at 4.22 K). (To obtain still lower temperatures, as low as 0.0014 K, a process known as adiabatic demagnetization is used, again starting at the temperature of liquid helium.)

The high energy requirements for compressing gases should not be regarded as all bad. Gases provide a medium with a high degree of interconvertibility between mechanical energy and thermal energy, and this is essential for practical power cycles and refrigeration processes.

Compression of ideal gases has already been treated in Chapter 5 (see the last part of that chapter, Examples 5.1 and 5.2, and the problems). There are several approaches to the compression of real gases. Examples 11.3 and 11.4 reworked Examples 5.1 and 5.2 with the use of the steam tables as a source of detailed thermodynamic data. An equivalent method would utilize graphical presentations, such as a *p–H*, *T–S*, or *H–S* diagram for the particular substance involved. Many common materials of industrial importance have been investigated extensively, and such tables and diagrams are available. The computer can be used to generate detailed tables that make interpolation easy and are convenient when the same substance is involved in a number of calculations.

For moderate pressures, the virial equation provides a means for correcting for departures from the ideal-gas behavior. For adiabatic compression in steady-flow processes, Eq. 4.34 yields

---

[1]Exceptions are absorption refrigeration, where the energy input comes from burning natural gas, and thermoelectric refrigeration, where there are no moving parts and no gas compression.

$$dH = -dW_s = C_p dT + \left(\frac{\partial H}{\partial p}\right)_T dp. \tag{12.8}$$

For an ideal gas, $(\partial H/\partial p)_T$ is equal to zero. Equations 9.11 and 9.9 show how to obtain this quantity as well as the pressure correction to C_p on the basis of the virial equation. For a reversible, adiabatic compression, Eq. 5.26 suggests that

$$dW_s = -Vdp, \tag{12.9}$$

but Eq. 7.29 still needs to be integrated with dS set equal to zero in order to determine the path of temperature and pressure that is followed during the compression. Analysis of this isentropic path is usually the first step also when the energy efficiency of the compressor is specified relative to the reversible, adiabatic work.

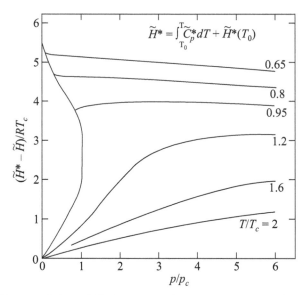

Figure 12.4 Enthalpy of fluids expressed as a correction to the value for an ideal gas. While the data are for nitrogen [1], they are plotted in a form made dimensionless with the critical properties so that they can be applied to any corresponding-states fluid. $\tilde{H}^*(T_0)$ is a constant reflecting the arbitrarily selected zero point for the enthalpy.

One last way to treat real gases is to correlate departures from ideal-gas behavior by means of the critical properties according to the corresponding-states principle of Chapter 3, Section 3.5. This is done in Fig. 12.4 for the enthalpy and Fig. 12.5 for the entropy. Figure 12.4 is very similar to a *p–H* diagram like Fig. 11.10, except that the integral of the ideal-gas heat capacity C_p^* is subtracted from the enthalpy. Greater generality is thereby achieved because the ideal-gas heat capacity depends on different factors not closely related to the PVT properties, as discussed in Chapter 5, Section 5.1. On the other hand, the variation of both the enthalpy and the entropy at constant temperature is determined entirely by the PVT properties, as shown by Eqs. 7.28 and 7.35. Thus, a corresponding-states correlation of the compressibility factor can be differentiated and integrated to yield corrections to the ideal-gas enthalpy and entropy. The corresponding deviation of the Gibbs function from the ideal-gas behavior is shown for another purpose in Fig. 18.1.

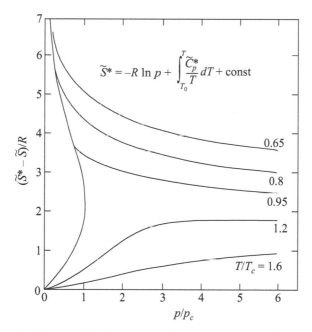

Figure 12.5 Entropy of fluids expressed as a correction to the value for an ideal gas, as in Fig. 12.4 for the enthalpy. Note that $\tilde{S}^*$ depends on the pressure as well as the temperature, and the constant in $\tilde{S}^*$ reflects the arbitrary zero point for the entropy.

Figures 12.4 and 12.5 can be applied to substances for which only incomplete data are available. The actual use is more awkward than the use of complete tables or diagrams. It is important to consider how all these complications could be avoided for ideal gases. Equation 5.9 seems to be the key to simplicity, allowing ideal-gas compression to be discussed before introduction of the second law of thermodynamics.

Example 12.1 Two identical bodies, each having the equation of state

$$U = CT,$$

are isolated as in Fig. 12.6 (compare Example 11.2). How much work can be extracted from these bodies, and what will be the final temperature?

Solution: We work this problem in two ways. First, regard a differential Carnot engine to operate, using the two bodies as the hot and cold reservoirs. We know that a Carnot engine will produce the maximum work for any heat engine operating between two reservoirs at given temperatures; here the temperatures of the reservoirs will change during the process.

Let the heat leaving Body 1 be Q_1 and that leaving Body 2 be Q_2. Energy balances for the two bodies yield

$$-dQ_1 = dU_1 = CdT_1$$

and

$$-dQ_2 = dU_2 = CdT_2,$$

and an energy balance for the engine is

$$dW = dQ_1 + dQ_2,$$

since the engine operates in a cyclic manner. For a Carnot engine, the heats are related to the temperatures according to

$$\frac{dQ_1}{T_1} + \frac{dQ_2}{T_2} = 0.$$

Elimination of dQ_1 and dQ_2 gives

$$\frac{dT_1}{T_1} + \frac{dT_2}{T_2} = 0$$

and

$$dW = -C(dT_1 + dT_2).$$

Integration of the first of these equations yields

$$\ln T_1 + \ln T_2 = \text{contant} = \ln T_1^o + \ln T_2^o$$

or

$$T_2 = \frac{T_1^o T_2^o}{T_1},$$

where T_1^o and T_2^o are the initial temperatures of bodies 1 and 2. The second differential equation becomes

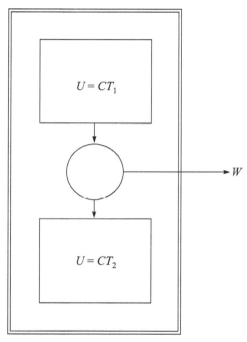

Figure 12.6 Two identical bodies, initially at different temperatures, are allowed to equilibrate through a heat engine that produces the work W.

$$dW = C\left(\frac{T_2}{T_1} - 1\right)dT_1 = C\left(\frac{T_1^o T_2^o}{T_1^2} - 1\right)dT_1.$$

From this equation, it is evident that no more work is produced when $T_1 = T_2$ and that the final temperature is the geometric mean of the initial temperatures

$$T_1 = T_2 = \sqrt{T_1^o T_2^o}.$$

Integration of the differential equation yields the work

$$W = C \int_{T_1^o}^{\sqrt{T_1^o T_2^o}} \left(\frac{T_1^o T_2^o}{T_1^2} - 1 \right) dT_1 = C(T_1^o + T_2^o - 2\sqrt{T_1^o T_2^o}).$$

Another way to work the problem is to realize that at equilibrium, the final temperatures will be equal and that, since a maximum-work process is reversible, the entropy change of the universe will be zero. Since a work source involves no entropy change, we have

$$\Delta S_{universe} = C \ln \frac{T_1}{T_1^o} + C \ln \frac{T_2}{T_2^o} = 0.$$

Since $T_1 = T_2$ in the equilibrium state, this gives

$$T_1 = T_2 = \sqrt{T_1^o T_2^o} .$$

An energy balance yields the work:

$$W = -\Delta U_1 - \Delta U_2 = -C(T_1 - T_1^o + T_2 - T_2^o)$$
$$= C(T_1^o + T_2^o - 2\sqrt{T_1^o T_2^o}).$$

Since work is produced in this process, the final temperature is lower in this maximum-work process than in Example 11.2, where no work was produced. We thus have a thermodynamic proof that the geometric mean is less than the arithmetic mean:

$$\sqrt{T_1^o T_2^o} < \frac{T_1^o + T_2^o}{2} .$$

Example 12.2 Calculate the work required and the temperature rise when water, initially at 79.58°F and 0.50 psia, is pumped reversibly and adiabatically to a pressure of 2000 psia and into the preheater of the steam power plant in Section 12.1 of the chapter.

Solution: Equation 7.29 with $dS = 0$ gives the relationship between the temperature and pressure for a reversible, adiabatic (isentropic) process

$$dT = \frac{T}{C_p} \left(\frac{\partial V}{\partial T} \right)_p dp = \frac{T \hat{V} \beta}{\hat{C}_p} dp ,$$

where the definition of the coefficient of thermal expansion has been introduced. On the right side, we take $T = 79.58 + 459.67 = 539.25°R$, $\hat{V} = 0.01608$ ft^3/lb (from the table for the steam power plant),

$\hat{C}_p$ = 0.99883 cal/g-K = 0.99883 Btu/lb-°R, and β = 2.07 × 10⁻⁴/1.8
$(°R)^{-1}$ (from Example 7.1). Noting the conversion factor that 1 Btu =
778.17 ft-lb, we obtain

$$\Delta T = \frac{539.25 \times 0.01608 \times 2.07 \times 10^{-4}}{1.8 \times 0.99883 \times 778.17}(2000 - 0.5)144 = 0.369°R.$$

This is the basis for the temperature rise in the pump, shown in the
table for the steam power plant. The reversible shaft work follows
from Eq. 12.9:

$$\widehat{W}_s = -\hat{V}\Delta p = -\frac{0.01608 \times (2000 - 0.5)144}{778.17} = -5.95\frac{\text{Btu}}{\text{lb}}.$$

Example 12.3 A practical consequence of the pressure dependence
of the heat capacity is a temperature mismatch in heat exchangers
for cryogenic applications; the temperature difference at the cold
end is greater than the temperature difference at the hot end. How
does this arise?

Solution: A cryogenic plant conserves cooling costs by cooling
an incoming stream with an outgoing stream, as sketched in Fig.
12.7. Representative temperature profiles along the length of the
heat exchanger are indicated. In a practical situation, the incoming
stream might be air, the process carried out at the cold end might
be the distillation of partially liquefied air, and the outgoing streams
would then be relatively pure nitrogen and oxygen. In Fig. 12.7, we
have simplified this to a nitrogen system with an unspecified process
at the cold end.

In a countercurrent heat exchanger, the enthalpy change of the
hot stream with distance will equal the enthalpy change of the cold
stream (see Eq. 4.34). A plot of temperature versus enthalpy with
pressure as a parameter, as in Fig. 12.8, therefore, gives an indication
of the nature of the temperature profiles.

If the warm, high-pressure stream enters the heat exchanger at
296 K, the atmospheric-pressure stream enters at 82 K, and the total
heat exchange area has been selected so that the temperatures differ
by 3 K at the hot end, then the effluent temperature of the warm
stream and the temperature difference at the cold end will depend
on the pressure as follows:

p, atm	200	50	20	6
T_{H1}, K	160.6	141.7	115.9	96.6
ΔT_1, K	78.6	59.7	33.9	14.6
$\Delta \tilde{S}_{\text{universe}}$, J/mol-K	8.90	6.23	3.80	1.76

(At 6 atm, the warm stream penetrates slightly into the two-phase region.) The extent of irreversible behavior related to the transfer of heat from one stream to the other over a nonzero temperature difference can be assessed by the entropy increase of the universe for each mole of nitrogen going through the process. This is indicated in the table also.

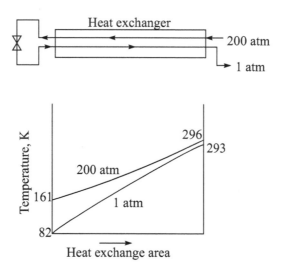

Figure 12.7 *Regenerative* heat exchange between two process streams, each having the same flow rate of nitrogen, but at different pressures. Unequal heat capacities of the two streams cause heat to be transferred across a substantial temperature difference over much of the length of the heat exchanger, and this is inherently an irreversible process.

Note that a longer heat exchanger can reduce the temperature difference at the hot end, but a significant difference will remain at the cold end. This represents an inherent irreversibility whenever such a heat exchanger is used in a process.

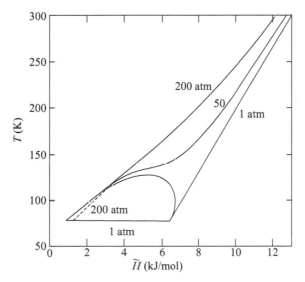

Figure 12.8 Temperature versus enthalpy for nitrogen. The slope of a curve of constant pressure on this plot is the reciprocal of the heat capacity $\tilde{C}_p$. Note that the isobar for 200 atm passes within the two-phase envelope. Of course the nitrogen never splits into two phases at a pressure greater than the critical pressure.

Over the range discussed here, the heat capacity of nitrogen is 56% higher at 200 atm than at 1 atm, and this information can be obtained from Fig. 12.4. That figure shows a larger enthalpy correction at lower temperatures, and this enthalpy correction must be *subtracted* from the ideal-gas value. At 200 atm, the reduced pressure of nitrogen is 5.963, and the reduced temperature is 2.344 and 1.272 at 296 and 160.6 K, respectively. From Fig. 12.4, $(\tilde{H}^* - \tilde{H})/RT_c$ is about 1 at the higher temperature and about 3 at the lower temperature. Assuming that the corrections are negligible at 1 atm, we estimate the change in the heat capacity to be

$$\tilde{C}_p - \tilde{C}_p^* = -\frac{RT_c}{\Delta T}(1-3) = \frac{8.3143 \times 126.26 \times 2}{296 - 160.6} = 15.5 \frac{\text{J}}{\text{mol-K}},$$

and this is about 53% of the ideal-gas value of $\tilde{C}_p^* = 3.5\,\text{R}$.

Example 12.4 Calculate the minimum amount of shaft work required in a steady-flow process for producing liquid nitrogen at 1 atm from gaseous nitrogen at 1 atm and 293 K.

Solution: The process can be depicted in Fig. 12.9, while Fig. 12.10 indicates a little more detail about how the cooling and liquefaction might actually be effected. Instead of analyzing a detailed process, we obtain more simply an expression for the minimum shaft work by applying the principle that the desired process is reversible and leaves the entropy of the universe unchanged. For this purpose, the temperature T_o of the reversible heat source must be clearly specified. Here we take $T_o = 293$ K, and the entropy change of the reversible heat source is then $-Q/T_o$, $-Q$ being the heat added to the reversible heat source. With a basis of 1 mol of nitrogen, we can write

$$\Delta S_{\text{universe}} = 0 = \tilde{S}_2 - \tilde{S}_1 - \tilde{Q}/T_o$$

while the energy balance from Eq. 4.34 becomes

$$\Delta \tilde{H} = \tilde{H}_2 - \tilde{H}_1 = \tilde{Q} - \tilde{W}_s .$$

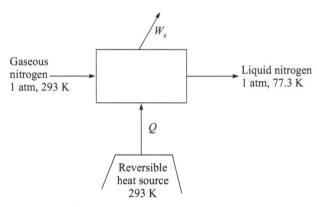

Figure 12.9 Steady-flow process to liquefy nitrogen.

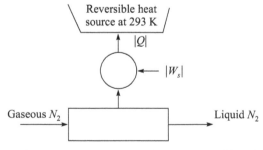

Figure 12.10 Schematic representation of how a reversible refrigerator (or a series of such refrigerators with various cold temperatures) might be used to liquefy nitrogen, with the consumption of shaft work.

Elimination of $\tilde{Q}$ yields the reversible shaft work

$$-\tilde{W}_s = \tilde{H}_2 - \tilde{H}_1 - T_o(\tilde{S}_2 - \tilde{S}_1).$$

The desired result could now be evaluated from thermodynamic properties taken from a table or diagram. Instead we shall estimate these by introducing a third state, saturated vapor at 77.3 K. Then $\Delta\tilde{H}_{vap}$ can be estimated as 5.3 RT_c from Fig. 12.4 and

$$\tilde{S}_3 - \tilde{S}_2 = \Delta\tilde{S}_{vap} = \Delta\tilde{H}_{vap} / T_2 = (\tilde{H}_3 - \tilde{H}_2)/T_2 .$$

Furthermore, the vapor phase can be treated as an ideal gas with a constant heat capacity of $\tilde{C}_p = 3.5\,R$. Thus,

$$\tilde{S}_1 - \tilde{S}_3 = \tilde{C}_p \ln (T_1/T_2)$$

and

$$\tilde{H}_1 - \tilde{H}_3 = \tilde{C}_p(T_1 - T_2).$$

The expression for the shaft work then becomes

$$-\tilde{W}_s = \Delta\tilde{H}_{vap}\left(\frac{T_o}{T_2} - 1\right) + \tilde{C}_p\left(T_2 - T_1 + T_o \ln \frac{T_1}{T_2}\right)$$

$$= 5.3 \times 8.3143 \times 126.26\left(\frac{293}{77.3} - 1\right)$$

$$+ 3.5 \times 8.3143\left(77.3 - 293 + 293 \ln \frac{293}{77.3}\right) = 2.06 \times 10^4\,\text{J/mol}.$$

This result can also be expressed as 0.204 kWh/kg of liquid nitrogen and agrees with the value calculated using the thermodynamic tables [1].

Problems

12.1 For this problem, state clearly which of the following methods you are using: steam tables, an *H–S* diagram for steam, or steam treated as an ideal gas. (In the last case, full credit will not be allowed unless you make corrections for nonideal behavior.)

   a. Steam enters an adiabatic turbine at 1000 psia and 1000°F and leaves at 100 psia. Calculate the maximum amount of work that can be obtained from the turbine, per pound of steam flowing.

    b. Steam enters an adiabatic throttling valve at 1000 psia and 1000°F and leaves at 100 psia. (The shaft work from a throttling valve is zero.) Calculate the "lost work" for this process, per pound of steam flowing. "Lost work" is equal to $T_0 \Delta S_{universe}$ (see Chapter 8), where T_0 is the absolute temperature of the ultimate heat sink that is accessible. Here, this temperature can be taken to be 60°F.

    c. Is the work produced in Problem (a) the same as the work lost in Problem (b), or is there any relationship between the two?

12.2 During a process, the thermodynamic system is constrained to go from state A to state B. The surroundings consist of a reversible heat source at a temperature T_0 and a reversible work source. Compare a reversible process and an irreversible process with regard to

    a. The internal energy changes of the system, heat source, and work source

    b. The entropy changes of the system, heat source, and work source

    c. The heat transferred from the system to the heat source

    d. The work W done on the work source by the system

    In which case is W larger?

12.3 For the steam power plant in Figs. 12.1–12.3, eliminate the reheater and the second turbine and instead condense the steam at 80 psia. Then pump the saturated condensate back to the preheater and boiler at 2000 psia. What is the efficiency of this modified process? (The answer should be 25.1%.)

12.4 Optimize the boiler and condenser temperatures for a steam power plant to give the maximum overall efficiency W/Q_H. The single turbine operates adiabatically and reversibly. The steam is not to condense in the turbine, and the temperature must not exceed 1100°F at any point in the system. Cooling water is available at 55°F.

12.5 Suppose that the pump in the steam power plant in Section 12.1 requires 50% more shaft work than the reversible amount. What then will be the temperature rise if the process remains adiabatic? (The answer should be 3.35°F.)

12.6 Compare $\Delta \tilde{H}_{\text{vap}}$ for a van der Waals fluid from Table 11.1 and for a Redlich–Kwong fluid from Table 11.2 with the value for a corresponding-states fluid from Fig. 12.4. Do this at a relatively low reduced temperature such as $T/T_c = 0.65$. Is there any relationship between this comparison and the vapor–pressure curves in Fig. 11.7?

12.7 New laws on the cogeneration of electricity along with process heat permit one to sell excess electricity to the power company at its "avoided cost." A gas turbine used for this purpose is claimed to have a high efficiency. We want to assess the validity of this claim. Treat the gas as an ideal gas with a constant heat capacity of $\tilde{C}_p^* = 3.5R$, and assume that this gas is taken through the following cycle:

1–2. Gas, approximating the fuel and air, is compressed adiabatically from 300 K and 1 atm to 13.3 atm with staged compressors that, in the aggregate, require shaft work 25% greater than the reversible work.

2–3. The gas is heated at constant pressure to 1367 K. This approximates the burning of the fuel–air mixture.

3–4. The gas is expanded through an adiabatic turbine to 1 atm. The expansion produces 20% less shaft work than an adiabatic reversible expansion to the same final pressure.

4–1. The gas is cooled at constant pressure to the initial condition.

Analyze the cycle and determine the efficiency of the power plant. Comment on the cycle and its efficiency.

Notation

A Helmholtz free energy, J
C heat capacity, J/K
C_p heat capacity at constant pressure, J/K
G Gibbs function, J
H enthlpy, J
p pressure, N/m^2
Q heat, J
R universal gas constant, 8.3143 J/mol-K
      or 82.06 atm-cm^3/mol-K

S entropy, J/K
T thermodynamic temperature, K
U internal energy, J
V volume, m^3
W work, J
W' work, excluding pressure–volume work, J
W_s shaft work, J
β coefficient of thermal expansion, K^{-1}
η efficiency

Subscripts, superscript, and special symbols

c critical point
C cold
H hot
vap   vaporization
*   ideal-gas state
~   per mole
^   per unit mass

Reference

1. R. M. Gibbons and G. P. Kuebler, *Thermophysical and Transport Properties of Argon, Neon, Nitrogen, and Helium-4* (Wright-Patterson Air Force Base, Ohio: Air Force Materials Laboratory, 1968, AFML-TR-68-370).

Chapter 13

Refrigeration and Heat Pumps

Refrigeration is a good application of the thermodynamics of one-component systems.

Figure 13.1 is a sketch of a Carnot refrigeration cycle on a T–S diagram. This is shown to be operating partly in the two-phase region for reasons we develop shortly. The efficiency is

$$\eta = \frac{W}{Q_H} = 1 - \frac{T_C}{T_H}. \tag{13.1}$$

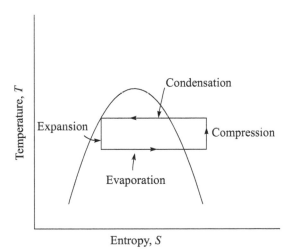

Figure 13.1 Carnot refrigeration cycle in the two-phase region.

The Newman Lectures on Thermodynamics
John Newman and Vincent Battaglia
Copyright © 2019 Jenny Stanford Publishing Pte. Ltd.
ISBN 978-981-4774-26-0 (Hardcover), 978-1-315-10861-2 (eBook)
www.jennystanford.com

The Carnot cycle has the lowest efficiency of any refrigerator operating between two reservoirs at given temperatures T_C and T_H, but it has the highest coefficient of performance (COP), where

$$\text{COP} = \left| \frac{Q_C}{W} \right| = \frac{1}{\eta} - 1. \tag{13.2}$$

The Carnot cycle would have the same COP if it were operated entirely in the vapor phase, for example, by lowering the pressures involved.

Figure 13.2 shows a practical refrigeration cycle that might be used in a typical small air conditioner, home refrigerator, or heat pump. (A heat pump operates on the same principle as a refrigerator, but its function is to heat a given space. The two ends of the unit in Fig. 13.2 are almost identical, but hot air comes out of one end, and cold air out of the other.) The compressor constitutes the only moving machinery in this system. In contrast, the reversible Carnot cycle in Fig. 13.1 requires machinery for all four parts of the cycle.

The evaporation of a pure fluid has the advantage that it can be carried out at constant temperature and at constant pressure simultaneously. Thus, one branch of the Carnot cycle can be an isothermal expansion involving evaporation in a simple heat exchanger with no moving parts.

The adiabatic compressor is hard to avoid. For practical reasons, it is not run in the two-phase region but only in the vapor region. In order to avoid more moving parts, the condenser is operated at constant pressure rather than constant temperature. This means that the vapor from the compressor is superheated, and this is no longer a Carnot cycle.

Finally, the adiabatic expansion is done with an irreversible throttling valve (Joule–Kelvin expansion) rather than a reversible turbine, as in the Carnot cycle, thereby eliminating more moving parts. In the power cycles of Chapter 12, this turbine can not, of course, be replaced since the work obtained from this device was the principal product.

The practical refrigeration cycle is sketched in a T–S diagram in Fig. 13.3. The entropy of the refrigerant is shown to increase in both the adiabatic expansion and the adiabatic compression, since these processes are now irreversible to some extent. The p–H diagram, Fig. 13.4, becomes a more "useful" representation of the refrigeration

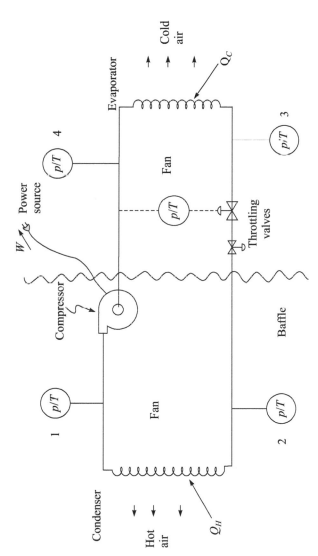

Figure 13.2 Schematic diagram of an air conditioner. All parts are to operate adiabatically except the heat exchangers used for the evaporator and the condenser.

cycle. Two parts of the cycle operate at constant pressure, and a third is at constant enthalpy. The shaft work in the compressor and the heat loads in the heat exchangers are now given directly by values of ΔH in the diagram (see Eq. 4.34).

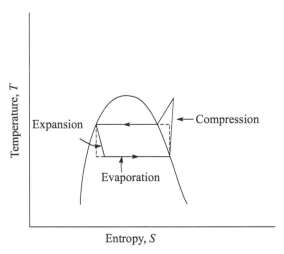

Figure 13.3 Practical refrigeration cycle on a *T–S* diagram. The dashed line indicates a Carnot cycle.

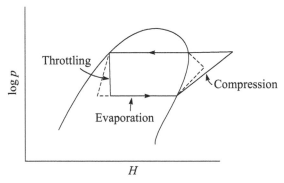

Figure 13.4 Practical refrigeration cycle on a *p–H* diagram. The dashed line indicates a Carnot cycle.

An important feature of operation in the two-phase region is that the throttling process does not significantly limit the capabilities of the system in comparison to a reversible, adiabatic expansion (which necessarily involves moving parts). The losses will be increased, but it is still possible to attain low temperatures without using multiple stages of refrigeration. To illustrate this point, Table 13.1 shows the temperature reached when several fluids are throttled from 100 psia to 35 psia, the initial temperature being 80°F in each case. Values of the temperature attained in a reversible, adiabatic expansion to the same final pressure are also given for comparison.

Table 13.1 Final temperatures attained when several fluids are expanded from 100 psia and 80°F to 35 psia by a throttling process ($\Delta H = 0$) and by a reversible, adiabatic process ($\Delta S = 0$).

	$\Delta H = 0$	$\Delta S = 0$
Freon 12 refrigerant	18°F	18°F
Ammonia	56	5
Freon 22 refrigerant	62	−4
Carbon dioxide	70	−30
Methane	~77	−42
Air	78.3	−55

In Table 13.1, Freon 12 is initially a liquid, and the final temperature is independent of whether $\Delta H = 0$ or $\Delta S = 0$ for the expansion, since this refrigerant is still in the two-phase region after the expansion. The Joule–Kelvin coefficient for air at 80°F was estimated in Example 9.3. Above the inversion curve, this coefficient is negative, and a Joule–Kelvin expansion would produce a somewhat warmer gas. The other substances are all vapors initially and in various positions relative to the critical point and the saturated vapor.

Two additional advantages of operating the refrigeration cycle in the two-phase region can be cited. The enthalpy of vaporization is moderately large, and consequently the amount of heat $\hat{Q}_C$ extracted per unit mass of refrigerant is normally larger than the value for a refrigerator operating entirely in the vapor region. Furthermore, the heat-transfer coefficients for evaporation and condensation of a pure fluid can easily be 100 times larger than for heat transfer to a vapor.

The design of the heat exchangers in Fig. 13.2 requires consideration of the transfer of heat from the refrigerant, through the exchanger wall, and to the air and is outside the scope of this text. Nevertheless, we should realize that we should allow for approximately 10°F between the refrigerant and the air in each of these exchangers.

The design of a refrigeration cycle usually requires us to take numbers for H, S, T, and p from a p–H diagram or from tables of the thermodynamic properties for the refrigerant. This is a good exercise for readers, and Example 13.2 illustrates the detailed analysis of an actual refrigeration cycle. With a little patience, one could also carry out the analysis using Figs. 12.4 and 12.5, which record thermodynamic properties for a corresponding-states fluid. Then one would also need to know the critical temperature and pressure and the ideal-gas heat capacity for the refrigerant (see Tables 3.1 and 5.1 and Fig. 5.1).

This detailed analysis may get a little tiresome, and we may seek a shortcut for the estimation of COP. Figure 13.5 provides a correlation of COP for an *ideal, standard* refrigeration cycle. By this we mean, with reference to Fig. 13.4, a cycle where the throttling begins with a saturated liquid, the evaporation continues only until we achieve a saturated vapor, and the compression begins at this point and is *reversible* and adiabatic. From the coefficient of performance of this ideal cycle, we can easily get COP for a real cycle by multiplying by the efficiency of the compressor (expressed as a ratio of reversible, adiabatic work to actual, adiabatic work for the same pressure ratio). Allowance for irreversible heat transfer would be made by selection of the condensation temperature a few degrees higher than that of the ambient heat sink and selection of the evaporation temperature a few degrees lower than required for the cold process stream.

The COP for the ideal cycle is divided in Fig. 13.5 by the value of the COP for a Carnot refrigerator operating between heat reservoirs at the condensation and evaporation temperatures. The fact that the ratio plotted in Fig. 13.5 is less than 1 then reflects irreversibilities associated with the Joule–Kelvin throttling and with removing the superheat of the refrigerant leaving the compressor. The reduced evaporation temperature can be obtained from the pressure ratio by means of a vapor–pressure curve (see Eq. 11.15 and Figs. 11.3 and 11.7).

We notice from Fig. 13.5 that the performance drops significantly for condensation temperatures approaching the critical temperature.

Because of the shape of the two-phase envelope in Fig. 13.4, the enthalpy of saturated liquid increases rapidly near the critical point, and this permits less heat to be absorbed at the lower-pressure evaporation. Furthermore, the reversible work of compression of saturated vapor by a given pressure ratio increases as the initial pressure increases. These effects are compensated to some extent by a larger ratio of condensation temperature to evaporation temperature.

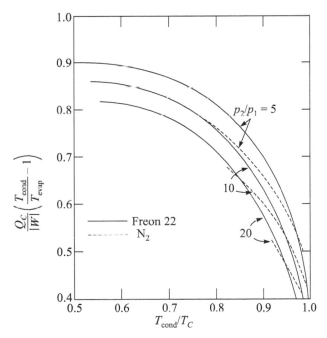

Figure 13.5 Coefficient of performance of *ideal, standard* refrigeration cycles relative to the coefficient for a reversible refrigerator operating between heat reservoirs at T_{cond} and T_{evap}. The correlation is expressed in terms of the reduced temperature of condensation and the pressure ratio between the condenser and the evaporator.

One might ask whether a representation like Fig. 13.5 should be able to correlate COP values for different refrigerants having different critical properties. Two factors have a bearing on this question. Fluids whose PVT properties obey the principle of corresponding states will have similar vapor pressure curves (Fig. 11.3) and similar enthalpies of evaporation at corresponding reduced temperatures (as given in Fig. 12.4). However, ideal-gas heat capacities enter into the work required for reversible, adiabatic compression—even when corrections are made for nonideal behavior. Consequently, two corresponding-states fluids will have somewhat different curves in Fig. 13.5 if the ideal-gas heat capacities are different. This effect is seen in Fig. 13.5 since $\tilde{C}_p^* = 3.5R$ for nitrogen while $\tilde{C}_p^*$ ranges from 5.3R to 7.03R for Freon 22 refrigerant for temperatures between −150°F and +150°F.

An important technological process closely related to refrigeration is the liquefaction of gases. Combinations of three principles are used for this purpose, and these involve:

1. Throttling or Joule–Kelvin expansion of the substance to be liquefied.
2. Expansion of the substance to be liquefied, accompanied by the low-temperature extraction of work approaching 80% of the theoretically possible work.
3. Cooling and condensation of the process fluid by a refrigeration system generally involving a different working fluid as the refrigerant.

Let us discuss these in reverse order. The practical refrigeration cycle already treated in Figs. 13.2 and 13.4 can be reasonably effective in extracting heat from a cold reservoir and delivering it to a hot reservoir. However, a single refrigeration system operates over only a limited temperature range. For example, in constructing Fig. 13.5, the largest temperature ratio encountered was T_{cond}/T_{evap} = 1.53 for a pressure ratio of 20. But in liquefying nitrogen, we are interested in a temperature ratio of 3.88.

A solution to this problem is *cascade refrigeration* with a sequence of refrigerants of successively lower critical temperatures. In Example 13.3, we make estimates for the work required to liquefy nitrogen by means of a cascade using ammonia, ethylene, methane, and nitrogen as refrigerants. Davies [1] estimates the requirement at 0.54 kWh of work for each kilogram of nitrogen produced as a liquid (compare Example 12.4). For the system he analyzes, the compressor for each refrigerant is operated at ambient temperature and above, with the compressed gas being cooled back to ambient temperature before entering the regenerative heat exchangers. This configuration is evidently superior energetically to the straightforward cascade treated in Example 13.3. In addition, the compressors are not required to operate at cryogenic temperatures where lubrication and metal fracture are problems.

Despite the superior energy requirements, cascade refrigeration is not widely used to liquefy air or oxygen and nitrogen, although it may be used for methane to produce liquefied natural gas (LNG). (A system developed for LNG actually involves a mixture of refrigerants —perhaps methane, ethane, and propane—which are all compressed together, followed by a series of Joule–Kelvin expansions.) For air, the complication of the refrigeration system may be the dominant factor favoring the processes that use air itself as the working fluid to produce low temperatures.

Table 13.1 dramatizes the desirability of extracting work during an expansion process if the objective is to achieve a low temperature, and this is particularly true for gases. The Claude process, which became commercial in 1902, utilized expansion of air in a cylinder-piston device from a temperature of perhaps 200 K and a pressure in the range from 20 to 200 atm. After expansion, the air attained 110 K at 1 atm and was used to cool another stream of high-pressure air. Thus, liquefaction was avoided in the cylinder expansion itself; it was accomplished by subsequent Joule–Kelvin expansion of the cold, high-pressure stream. The work requirement ranged from 1.23 to 0.89 kWh/kg of liquid air as the high pressure ranged from 20 to 100 atm. Whether the work extracted in the cylinder expansion is utilized in the compression or is merely dissipated is of minor importance; the main feature is that this amount of energy is removed from the stream to be liquefied.

Kapitza developed a turbine expander capable of extracting 80% of the work possible in an isentropic expansion. His air liquefaction system could then operate with air compressed initially to only about 6 atm, although the work requirement might remain at about 1.1 kWh/kg of liquid air.

The simple Linde process, which produced the first commercial liquefaction of air (1895), consisted of the Joule–Kelvin expansion of air from 200 atm to 1 atm after cooling with the exit vapor in a regenerative heat exchanger. Figure 12.7 could represent this system if we include provision for removing liquid at the cold end after the air passes through an adiabatic throttling valve. The work requirement can be reduced from about 2.5 to 1.4 kWh/kg by using instead a two-stage expansion with an intermediate pressure of 20 to 50 atm.

The importance of attaining a low temperature before the Joule–Kelvin expansion can be appreciated from the *T–H* diagram in Fig. 12.8. Air at 200 atm must be cooled to 170 K before any liquid will be produced by the expansion; at 50 atm, the corresponding temperature is about 140 K. Hydrogen has a particularly low inversion temperature (see problem 9.2); it must be cooled to about 202 K before a Joule–Kelvin expansion will produce any cooling at all, much less any liquid.

Other modifications of the simple Linde process to make it more energy efficient include use of a turbine expander somewhere in the system and the initial cooling of the high-pressure air to about 220 or 240 K by means of an ammonia refrigerator. The auxiliary refrigerator is the first step toward a cascade refrigeration system. With the use of the turbine expander, the modified Linde process involves all three methods for attaining low temperatures: refrigeration, Joule–Kelvin expansion, and expansion with the extraction of work.

A summary of energy requirements for liquefaction of nitrogen (or air) is given below.

Reversible	0.204 kWh/kg
Refrigeration cascade	0.54
Claude process (cylinder expansion)	0.89
Kapitza process (turbine expansion)	1.1
Two-stage Linde with precooling to 228 K	0.89
Simple Linde with precooling to 228 K	1.5
Two-stage Linde	1.4
Simple Linde process	2.5

The usefulness of thermodynamic analysis can be seen in the evaluation of process alternatives for refrigeration and liquefaction processes. The maximum efficiency that could possibly be attained can quickly be established. Then several flow sheets that compromise efficiency for simplicity and reliability to various degrees can be devised and compared. Even though thermodynamics cannot give a thorough analysis of irreversible processes, it can be used to assess the entropy production for some commonly encountered irreversibilities such as the limited efficiency of compressors and expansion machines, the superheat produced in a compressor, the temperature difference required for heat transfer (including the mismatch of the heat capacities of two process streams), and the loss in the Joule–Kelvin throttling expansion.

Example 13.1 A fan blows air of constant heat capacity across the condenser of a refrigerator system, the air being heated from 65°F to 70°F (see Fig. 13.6). Within the coils, ammonia is cooled and condensed at 150 psia from 187°F to the condition of saturated liquid.

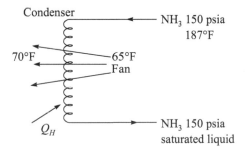

Figure 13.6 Condenser in an ammonia refrigeration loop.

Calculate the entropy change of the universe. Express your answer in Btu/°R per pound of ammonia circulating in the refrigeration loop. You may neglect the work consumed by the fan.

Solution: Let f be the flow rate of air. Then

$$\Delta S_{air} = f\,\tilde{C}_p \ln \frac{T_2}{T_1},$$

where T_2 = 70°F = 529.67°R and T_1 = 65°F = 524.67°R. An energy balance yields

$$Q_H = -f\tilde{C}_p(T_2 - T_1) = \Delta \widehat{H}_{NH_3}.$$

Hence (here the circumflex means per pound of ammonia)

$$\Delta \widehat{S}_{air} = -\Delta \widehat{H}_{NH_3} \frac{\ln T_2/T_1}{T_2 - T_1} = -\frac{\Delta \widehat{H}_{NH_3}}{527.186°R}.$$

From the p–H plot for ammonia,

$$\Delta \widehat{H}_{NH_3} = 130 - 701 = -571 \text{ Btu/lb},$$

and $\quad \Delta \widehat{S}_{NH_3} = 0.27311 - 1.32 = -1.04689 \text{ Btu/lb-°R}.$

($\hat{S}_{liq}$ must be obtained from $\hat{S}_{vap} = 1.205\,\text{Btu/lb-}^\circ\text{R}$ and $\Delta\hat{S}_{vap} = \Delta\hat{H}_{vap}/T_{sat} = 502/538.67 = 0.93189\,\text{Btu/lb-}^\circ\text{R}$ at the condensation temperature of 79°F.)

Consequently, the entropy change of the universe is

$$\Delta\hat{S}_{universe} = \Delta\hat{S}_{air} + \Delta\hat{S}_{NH_3}$$

$$= \frac{571}{527.186} - 1.04689 = 0.0362\frac{\text{Btu}}{\text{lb-}^\circ\text{R}}.$$

Some care is required in order not to lose all the significant figures when subtracting nearly equal numbers.

Example 13.2 The air conditioner in Fig. 13.2, utilizing Freon 22 refrigerant, consumes 752 W of electric power in the compressor and yields the following data (taken by Douglas N. Bennion at U.C.L.A. in November, 1971) for the pressure and temperature at various points in the cycle:

S. No.	Location	Temperature (°F)	Pressure (psia)
1.	After compressor	177	257
2.	After condenser	100	251
3.	After throttling valves	40	83
4.	After evaporator	41	82

The air conditioner sits open in a room where the temperature T_o is 69°F. With the heat capacity of air taken to be $\tilde{C}_p = 6.92\,\text{cal/mol-K}$, integration of values of the air velocity and the air temperature (measured at 54 points downstream of the condenser and at 54 points downstream of the evaporator) yields a heat load of $Q_C = 144{,}992$ J/min and an average downstream air temperature of $T_C = 47.7$°F for the evaporator and a heat load of $Q_H = -187{,}429$ J/min and an average downstream air temperature of $T_H = 95.9$°F for the condenser.

The heat losses from the compressor and motor were estimated to be 2899 J/min. In addition, we are told that the compressor is single acting with a single cylinder having a displacement of 0.782 in^3. It operates at 3450 rpm under normal load, has a stall speed of 3200 rpm, and operates at 3500 rpm at light load.

From this information, we should like to analyze the behavior of the air conditioner by finding the fraction of liquid in the stream after the expansion valves, performing energy balances and estimating the rate of circulation of the refrigerant, calculating the COP of the device, and assessing where the entropy production occurs.

Solution: From the temperatures and pressures recorded above, the thermodynamic tables for the refrigerant yield the volume, enthalpy, and entropy at various points in the system:

		$\hat{V}$ (ft³/lb)	$\hat{H}$ (Btu/lb)	$\hat{S}$ (Btu/lb-°R)
1.	After compressor	0.25869	126.45	0.2315
2.	After condenser	0.01404	39.37	0.07942
3.	After throttling valves	0.1461	39.37	0.08225
4.	After evaporator	0.6707	108.39	0.2207
	Vapor at 3	0.6575	108.14	0.2199
	Liquid at 3	0.01262	21.42	0.04632

In looking up these data, we notice that the liquid at 2 is sub-cooled by 13°F at the exit of the condenser, and the vapor at 4 is superheated by 2°F before entering the compressor (compare Problem 13.2). The fluid at 3 after the throttling valves is a mixture of liquid and vapor at 40°F. We determine the fraction x of liquid by assuming that the enthalpy of this stream is unchanged from that at point 2 (before the throttling valves):

$$\hat{H}_3 = x\hat{H}_{\text{liq}} + (1-x)\hat{H}_{\text{vap}}$$

or

$$x = \frac{39.37 - 108.14}{21.42 - 108.14} = 0.793,$$

and from this value of x, we calculate the volume and entropy of the refrigerant at point 3.

An overall energy balance yields

$$W = Q = 144{,}992 - 187{,}429 - 2{,}899 = -45{,}336 \text{ J/min} = -756 \text{ W},$$

in good agreement with the measured value of 752 W.

The air conditioner does not have a flow meter, which measures directly the circulation rate of the refrigerant. In the absence of such information, the compressor can act as a flow meter (see Examples 5.1 and 11.3). An energy balance on the compressor, with due allowance for the estimated heat loss, yields

$$F = \frac{752 \times 60 - 2899}{(126.45 - 108.39) \times 1054.8} = 2.216 \frac{\text{lb}}{\text{min}},$$

there being 1054.8 J/Btu.

Energy balances on the evaporator and on the condenser yield two additional estimates of the circulation rate of the refrigerant:

$$F = \frac{144,992}{(108.39 - 39.37) \times 1054.8} = 1.992 \frac{\text{lb}}{\text{min}}$$

and

$$F = \frac{187,429}{(126.45 - 39.37) \times 1054.8} = 2.041 \frac{\text{lb}}{\text{min}}.$$

The speed and displacement of the compressor and the volume $\hat{V}_4$ allow a fourth estimate: $F = 2.328$ lb/min. The average of these four values is 2.144 lb/min, with a deviation of 6.5%. Notice that the energy balances on the branches of the cycle are not as consistent as we might expect from the good agreement found in the overall energy balance.

The COP of the refrigerator is

$$\text{COP} = -\frac{Q_C}{W} = \frac{144,992}{752 \times 60} = 3.22.$$

This can be compared with the value COP = 10.53 for a Carnot refrigerator operating between reservoirs at the temperatures 47.7°F and 95.9°F and the value COP = 6.74 for a Carnot refrigerator operating between the condensation temperature of 113°F and the evaporation temperature of 39°F.

The rate of entropy production in the various parts of the system is given below:

	Entropy production rate (Btu/°R-min)	Percent of total
Compressor	0.029135	36.0
Condenser	0.01902	23.5
Throttling valves	0.006272	7.8
Evaporator	0.01151	14.2
Mixing of air in room	0.014975	18.5
Total (for universe)	**0.080912**	**100**

The entropy production rates in the condenser and in the evaporator can be calculated along the lines indicated in Example 13.1. For the evaporator, this rate is

$$F\left[\hat{S}_4 - \hat{S}_3 - \frac{\hat{H}_4 - \hat{H}_3}{T_C - T_o}\ln\frac{T_C}{T_o}\right]$$

$$= 2.216\left[0.2207 - 0.08225 - \frac{108.39 - 39.37}{47.7 - 69}\ln\frac{47.7 + 459.67}{69 + 459.67}\right]$$

$$= 2.216 \times (0.13845 - 0.133257) = 2.216 \times 0.005193 \text{ Btu/°R-min},$$

and for the condenser we have

$$F\left[\hat{S}_2 - \hat{S}_1 - \frac{\hat{H}_2 - \hat{H}_1}{T_H - T_o}\ln\frac{T_H}{T_o}\right]$$

$$= 2.216\left[0.07942 - 0.2315 - \frac{39.37 - 126.45}{95.9 - 69}\ln\frac{95.9 + 459.67}{69 + 459.67}\right]$$

$$= 2.216 \times (-0.15208 + 0.16066) = 2.216 \times 0.0085816 \text{ Btu/°R-min}.$$

The two nearly adiabatic units, the expansion valves and the compressor, each show an increase in entropy of the refrigerant. The rate of entropy production in the valves is

$$F(\hat{S}_3 - \hat{S}_2) = 2.216 \times (0.08225 - 0.07942) \text{ Btu/°R-min, and}$$

that for the compressor is

$$F(\hat{S}_1 - \hat{S}_4) + \frac{Q_{comp}}{T_o} = 2.216 \times (0.2315 - 0.2207) + \frac{2899}{528.67 \times 1054.8}.$$

The entire process takes the electrical energy $-W$ and converts it into internal energy at the room temperature T_o. Hence, the entropy production for the universe is

$$\Delta S_{universe} = \frac{752 \times 60}{528.67 \times 1054.8} = 0.080912\frac{\text{Btu}}{\text{min-°R}}.$$

The remainder of the entropy is produced outside the air conditioner by the mixing of the hot and cold air streams with the air in the room.

These calculations require some precision in order to obtain meaningful results. This is particularly true for the condenser and the evaporator, where the entropy production is small compared to the entropy transferred between the air and the refrigerant. One should notice that the throttling valves are not responsible for a large fraction of the entropy production, and there is little incentive to replace these simple devices with more efficient turbines.

The refrigeration cycle is plotted on an *H–S* diagram in Fig. 13.7. Readers can plot this cycle in the *p–H* diagram in Appendix if they wish, or they can construct *p–V* or *T–S* diagrams to illustrate the cycle.

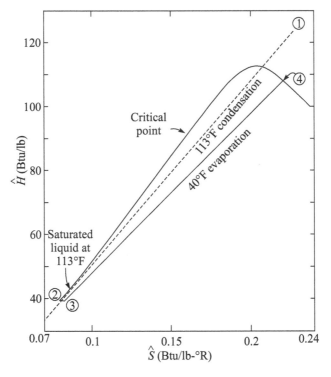

Figure 13.7 Refrigeration cycle of Example 13.2 on an *H–S* diagram for Freon 22 refrigerant. On this plot, a curve of constant pressure has a slope equal to *T*.

Example 13.3 Assess the energy requirements for cascade refrigeration to liquefy nitrogen, where the evaporation pressures are all chosen to be atmospheric. (Subatmospheric pressures would increase the leakage problems and increase the volume of gas to be compressed.) Each compressor requires 25% more shaft work than that corresponding to an isentropic compression over the specified pressure ratio. The evaporation and condensation conditions are specified as follows:

	Evaporation		Condensation	
	T, K	*p*, atm	*T*, K	*p*, atm
Ammonia	239	1	298	10.2
Ethylene	169	1	242	19
Methane	112	1	172	24.7
Nitrogen	77.3	1	114.6	18.73

Solution: A flow sheet for the system is shown in Fig. 13.8. Condensation temperatures are evidently based on the necessity for an adequate temperature difference for heat transfer to the adjacent evaporator in the cascade. The resulting pressure ratio for each refrigerant is substantial.

The heat load (Q_C) on, for example, the ethylene refrigerator is the sum of the heat load and the work for the methane refrigerator plus the cooling requirement for the nitrogen in the ethylene bath (2.045 kJ in Fig. 13.8) as the nitrogen is being cooled from ambient temperature to the condensation temperature. On the basis of 1 mol of nitrogen to be liquefied, these energy flows are indicated in Fig. 13.8.

Without assembling tables and diagrams for the four refrigerants, an approximate analysis can be based on the correlation in Fig. 13.5. Again for ethylene, the reduced condensation temperature is 242/283.05 = 0.8550, and the pressure ratio is 19. From Fig. 13.5, we read a value of 0.655 (use the curves for Freon 22 for all the refrigerants except nitrogen). Multiply this value by 2.315, the COP for a Carnot refrigerator, and by 0.8, the compressor efficiency, to yield an estimated COP = 1.213 for the ethylene refrigerator. Similar estimates are made for the other refrigerators, and these are recorded in Fig. 13.8. The total work obtained in this way is 83.7 kJ or 0.83 kWh of work required to liquefy 1 kg of nitrogen by the refrigeration cascade shown in Fig. 13.8.

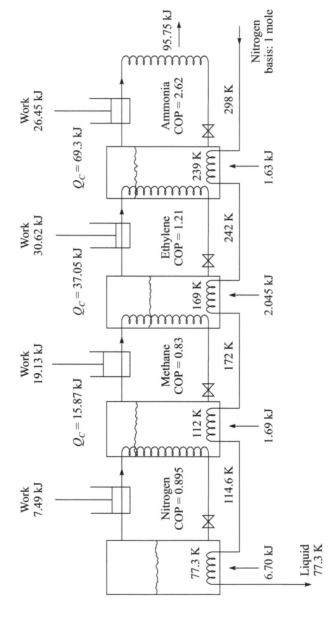

Figure 13.8 Refrigeration cascade for liquefying nitrogen.

Problems

13.1 The heat losses from a house are 10 kW, and electricity costs 4.2 ¢/kWh. The house is at 25°C, and the outside air is at 0°C.

  a. How much would it cost per hour to heat the house with an electric resistance heater?

  b. How much would it cost per hour to heat the house with a reversible heat pump operating between the outside air and the air in the house?

13.2 The throttling valve with the controller in Fig. 13.2 is intended to ensure that the fluid entering the compressor is slightly superheated, since liquid present in the compressor would promote erosion of the moving parts. Discuss the possibility of condensation within the compressor if the inlet stream has only one degree of superheat. Compare with the situation in a turbine in the power cycles of Chapter 12.

13.3 A refrigerator using pure ammonia as a refrigerant involves the following cycle:

  a. Irreversible expansion of saturated liquid through an adiabatic throttling valve from 150 psi to 35 psi.

  b. Evaporation of the remaining liquid at 35 psi. The refrigerant is in contact with the cold reservoir during this part of the cycle.

  c. Reversible adiabatic compression to the original pressure.

  d. Cooling and condensation at this pressure until all the refrigerant is in the form of saturated liquid.

  What is the highest temperature reached by the refrigerant in this process, and at what point in the cycle does it occur?

13.4 For the refrigeration cycle described in Problem 13.3, what is the efficiency of the cycle? What is the efficiency of a Carnot cycle operating between the temperatures of the saturated liquid at 150 and 35 psi? Calculate the corresponding coefficients of performance for these two situations.

13.5 The heat leak Q_C into a cold storage room amounts to 15,000 Btu/h when maintained at 15°F under ambient conditions of 85°F (see Fig. 13.9). Design a refrigeration cycle by selecting a refrigerant and choosing operating conditions such that

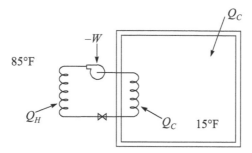

Figure 13.9 Cold storage room and refrigeration cycle.

    a. The condensation temperature is 10°F greater than ambient.

    b. The evaporation temperature is 10°F lower than the temperature of the cold storage room.

    c. The compressor operates adiabatically and should be assumed to consume 20% more power than a reversible, adiabatic compression from the evaporation pressure to the condensation pressure.

    d. A throttling (Joule–Kelvin) expansion is used to achieve the low temperature.

Calculate the circulation rate of the refrigerant, the volumetric flow rate at the input to the compressor, the COP, and the COP of one or more relevant Carnot refrigeration cycles. Analyze the entropy production rates in the system and its surroundings.

13.6 A refrigeration machine requires 1 kW of power per ton of refrigeration. (One ton of refrigeration is equivalent to a Q_C of 12,000 Btu/h or 3510 J/s.)

    a. What is the coefficient of performance?

    b. How much heat is rejected to the condenser?

    c. If the condenser operates at 60°F, what is the lowest temperature the refrigerator could possibly maintain?

13.7 For a system using Freon 22 refrigerant, it is decided to maintain an upper condensation temperature of 555.9°R (at 200 psia) and a lower evaporator temperature of 396.3°R (at 8 psia). Since this pressure ratio of 25 is considered large, it is desired to compare a single refrigeration cycle with a cascade consisting of two refrigerators in series, where condensation for the low-temperature cycle and evaporation for the high-temperature cycle take place at 461.3°R (at 40 psia). (Common sense may tell you that it is inefficient to have heat transfer between two streams of the same refrigerant at the same pressure. Other alternatives are explored in Problem 13.8.)

a. Compare COP for the single cycle with that for the two-cycle composite system where all cycles are assumed to be *ideal*, *standard* refrigeration cycles.

b. How is the comparison modified if each compressor now requires 25% more work than for a reversible, adiabatic compression?

c. How is the comparison modified if the composite system is also required to allow 5°F of overlap for heat transfer between the two cycles? Condensation in the low-temperature cycle now occurs at 463.8°R (at 42.1 psia), and evaporation in the high-temperature cycle now takes place at 458.8°R (at 38.0 psia).

13.8 The engineering application remains the same as in Problem 13.7, with Freon 22 refrigerant operating between an upper pressure of 200 psia and a lower pressure of 8 psia. Now it is desired to modify the single cycle to use a two-stage compression to an intermediate pressure yet to be determined. An interstage cooler can be used, but it can cool the refrigerant only to 555.9°R (mainly because the cooling water is only a few degrees below this value).

After the decision is made to use a two-stage compression, it is realized that a two-stage Joule–Kelvin expansion involves only a slight additional complication. The vapor from the first expansion enters the second stage of the compressor and is to be at the same pressure as the gas from the interstage cooler, which can now be further cooled by boiling some liquid from the first expansion. The flow sheet is now shown in Fig. 13.10. All compressor stages require 25% more work than the reversible, adiabatic work for the same pressure ratio. What is the COP for the single, unmodified cycle and for the modified cycle with several values for the intermediate pressure?

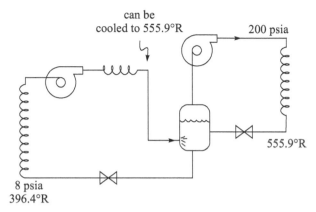

can be
cooled to 555.9°R   200 psia

8 psia
396.4°R

Figure 13.10 Refrigeration cycle with two-stage compression and two-stage expansion.

13.9 An ideal gas with a heat capacity of $\tilde{C}_p = 3.5R$ is used as a refrigerant in a system where the adiabatic compression and the isothermal compression involve pressure ratios of 2.6425 and 7.5687, respectively, and each requires 25% more work than the corresponding reversible value. The cycle also includes an adiabatic and an isothermal expansion, each of which yields 80% of the corresponding reversible work. Calculate the coefficient of performance for this refrigerator and compare it to the value for a Carnot refrigerator operating between the same hot and cold temperatures. Sketch the refrigeration cycle on a T–S diagram.

13.10 What is a realistic value to expect for the COP for a heat pump for space heating? The inside temperature is 70°F, the design temperature is 20°F for the outside ambient, and an extra 10°F is to be allowed for heat transfer at both the condenser and the evaporator. State the compressor efficiency you assume, the refrigerant, and its critical temperature and pressure.

13.11 The descriptions of the liquefaction processes may daze you. To clarify these in your mind, make sketches of flow sheets for the processes listed below. Be sure to include regenerative heat exchangers where appropriate.

   a. Claude process.

   b. Two-stage Linde process.

   c. Mixed refrigerant. (Hint: Remember that the low-boiling refrigerants remain in the vapor after a Joule–Kelvin expansion and must be forwarded toward the low-temperature end of the cascade.)

13.12 The performance of the cascade refrigerator in Example 13.3 can be improved by the use of economizers. Since multistage compression is needed anyway for the pressure ratios involved, there is little additional expense in adding a phase separator and a second throttling valve, as shown in Fig. 13.11. Redesign the part of the cascade using nitrogen as the refrigerant so that the intermediate pressure is the geometric mean of the evaporator and condenser pressures. Each compressor requires 25% more shaft work than that corresponding to a reversible, adiabatic compression over the specified pressure ratio.

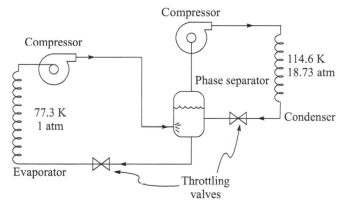

Figure 13.11 Refrigerator with two-stage throttling and compression. The vapor and liquid leaving the phase separator are in equilibrium. The refrigerant is N_2.

13.13 Based on the redesign of the nitrogen stage in the preceding problem, calculate the work required for nitrogen liquefaction by means of the modified cascade. If you get no result from the preceding problem, assume that the coefficient of performance of the nitrogen stage is improved from 0.895 to 1.5 by means of the design modification.

13.14 Saturated liquid chlorine at 38 atm is throttled in a Joule–Kelvin expansion to 7.6 atm. Estimate the initial and final temperatures and the fraction of liquid remaining after the expansion. Critical properties can be found in Table 3.1.

Notation

C_p	heat capacity at constant pressure, J/K
COP	coefficient of performance for a refrigerator
f	flow rate of air
F	flow rate of refrigerant, kg/s
H	enthalpy, J
p	pressure, N/m^2
Q	heat, J
R	universal gas constant, 8.3143 J/mol-K or 82.06 atm-cm^3/mol-K
S	entropy, J/K
T	thermodynamic temperature, K
T_0	room temperature, K
V	volume, m^3
W	work, J
x	fraction of liquid
η	efficiency of thermodynamic cycle

Subscripts and special symbols

c	critical point
C	cold
comp	compressor
cond	condensation
evap	evaporation
H	hot
liq	liquid
vap	vapor or vaporization
*	ideal-gas state
~	per mole
^	per unit mass

Reference

1. Mansel Davies, *The Physical Principles of Gas Liquefaction and Low Temperature Rectification* (London: Longmans, Green and Co., 1949), pp. 2–27. Energy requirements for other liquefaction methods are also taken from this source.

Part B

MULTICOMPONENT SYSTEMS

Chapter 14

Mixtures

Chemical engineers must be vitally concerned with mixtures, that is, systems containing more than one pure substance. This is because their principal objectives involve separation processes and chemical reactions, and these cannot occur in pure systems.

The extensions of thermodynamics to multicomponent systems were carried out principally by Gibbs [1]. This great American scientist also made important contributions to statistical mechanics, a field that gives much support to thermodynamics by establishing the relationship with molecular concepts.

A mixture can frequently be treated as though it were a pure substance. Air is perhaps the most common example. However, air must be regarded as a multicomponent system when one considers liquefaction, particularly for the purpose of separating air into its components, or humidification, or smog and chemical reactions in the atmosphere.

Furthermore, the laws of thermodynamics, developed in Part A, apply to a closed, multicomponent system just as well as to pure substances, even though there may be two phases present. Figure 14.1 depicts a pressure vessel for measurements in a calorimeter. This *bomb* contains water and ammonia, and these substances may be transferred from one phase to the other as the temperature is raised (compare Problem 3.2d). Nevertheless, as the temperature is

The Newman Lectures on Thermodynamics
John Newman and Vincent Battaglia
Copyright © 2019 Jenny Stanford Publishing Pte. Ltd.
ISBN 978-981-4774-26-0 (Hardcover), 978-1-315-10861-2 (eBook)
www.jennystanford.com

raised, the heat supplied to the contents of the bomb is still related to the entropy change according to

$$dQ_{rev} = TdS. \tag{14.1}$$

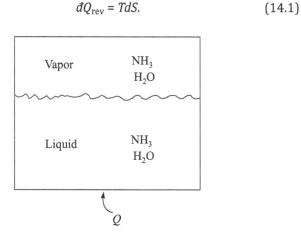

Figure 14.1 Calorimeter bomb for determining heat capacities.

In other words, the contents of the bomb can be used as the working material in a Carnot engine just as well as a pure substance. The Carnot engine can then be used to establish the thermodynamic temperature scale and the existence of a state property, known as entropy, for the working material of the engine.

The contents of the bomb, thus, comprise a PVT thermodynamic system characterized by internal energy U, entropy S, pressure p, volume V, temperature T, etc. The system behaves much like a pure substance, except that an isotherm in the two-phase region does not occur at constant pressure; see Fig. 14.2. A constant-pressure cooling brings a vapor mixture down to the *dew point*, the temperature at which liquid first appears. The temperature continues to drop during condensation until, at the *bubble point*, all the vapor is condensed and there is only liquid.

The calorimeter bomb in Fig. 14.1 is designed for the determination of the heat capacity of the *liquid* phase, but the vapor space is provided to keep the pressure from getting too high and bursting the bomb. In order to obtain the heat capacity $\hat{C}_p$ of the liquid from Eq. 14.1, one must make corrections for the change in volume of the inside of the bomb, for the evaporation or condensation of NH_3 and H_2O as the temperature is raised, for the fact that the pressure is not constant, and for the heat capacity of the vapor phase. For this purpose, we can no longer ignore the fact that this is a multicomponent system.

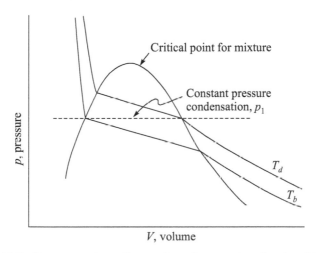

Figure 14.2 Pressure–volume diagram for a given number of moles of H_2O and NH_3. Illustrated are the isotherms for the dew-point temperature and bubble-point temperature for a given pressure p_1.

14.1 Partial Molar Quantities

In describing the properties of mixtures, the composition of the system needs to be included. For independent variables in the system of ammonia and water, we might take the internal energy U, the volume V, the number n_{NH_3} of moles of ammonia, and the number n_{H_2O} of moles of water. Other choices are also possible. U and V could be replaced by the temperature T and the pressure p; or the numbers of moles could be replaced by the total number of moles

$$n = n_{NH_3} + n_{H_2O} \qquad (14.2)$$

and the mole fraction of ammonia

$$x_{NH_3} = \frac{n_{NH_3}}{n}. \qquad (14.3)$$

For this two-component system, one more independent variable is needed than in the case of a pure substance.

We mention that the development of Part A applies to closed, multicomponent systems in order to emphasize that we already have the framework for describing variations at constant composition. Thus, we already know how to define, measure, and tabulate the isothermal compressibility

$$k = -\frac{1}{V}\left(\frac{\partial V}{\partial p}\right)_{T,n_j}, \tag{14.4}$$

the coefficient of thermal expansion

$$\beta = \frac{1}{V}\left(\frac{\partial V}{\partial T}\right)_{p,n_j}, \tag{14.5}$$

and the heat capacities

$$C_p = T\left(\frac{\partial S}{\partial T}\right)_{p,n_j} = \left(\frac{\partial H}{\partial T}\right)_{p,n_j} \text{ and } C_V = T\left(\frac{\partial S}{\partial T}\right)_{V,n_j} = \left(\frac{\partial U}{\partial T}\right)_{V,n_j}. \tag{14.6}$$

We need only remind ourselves that the numbers of moles of all the components *j* are to be kept constant during the indicated differentiations.

These quantities in Eqs. 14.4–14.6 cover the important PVT and thermal properties of the system. Important relationships in Chapter 7 still apply. For example, Eq. 7.32 relates the heat capacities, and Eqs. 7.7, 7.10, 7.13, 7.28, 7.33, and 7.35 describe the pressure or volume dependence of the entropy, the internal energy, the enthalpy, and the heat capacities.

We want to apply these relationships to single phases and to be in a position to predict or tabulate values of k, β, and $\hat{C}_p$ for single phases, rather than for composite systems like that depicted in Fig. 14.1. The principles of phase equilibria can then be developed separately after we know something about that properties of individual phases (see Chapter 11). Similarly, we are interested in the internal energy per mole $\tilde{U}$, or per unit mass $\hat{U}$, for single phases.

In addition to k, β, and $\hat{C}_p$, describing the pressure and temperature dependence, we need quantities describing the composition dependence of thermodynamic properties. Fundamental for this purpose are the partial molar quantities. For example, the partial molar volume is

$$\overline{V}_i = \left(\frac{\partial V}{\partial n_i} \right)_{T,p,n_j \atop j \neq i} , \qquad (14.7)$$

the partial molar enthalpy is

$$\overline{H}_i = \left(\frac{\partial H}{\partial n_i} \right)_{T,p,n_j \atop j \neq i} , \qquad (14.8)$$

and the partial molar Gibbs function is

$$\overline{G}_i = \left(\frac{\partial G}{\partial n_i} \right)_{T,p,n_j \atop j \neq i} . \qquad (14.9)$$

First, we notice that the composition here is represented by the numbers of moles of the components, not by the masses nor by the mole fractions or concentrations. Gibbs originally used masses rather than numbers of moles. The partial molar quantities are sufficient for describing the composition dependence; other representations can be obtained from them by suitable mathematical manipulations. For example, the derivative of the density of the ammonia–water system with respect to the ammonia concentration can be expressed as

$$\left(\frac{\partial \rho}{\partial c_{NH_3}} \right)_{T,p} = M_{NH_3} - M_{H_2O} \frac{\overline{V}_{NH_3}}{\overline{V}_{H_2O}} , \qquad (14.10)$$

where M_i is the molar mass of species i.

Second, we notice that the temperature and pressure are the quantities chosen to be held constant in defining the fundamental composition derivatives. This permits us to write down directly the volume in terms of the partial molar volumes

$$V = \sum_i n_i \overline{V}_i = n\tilde{V} , \qquad (14.11)$$

or the Gibbs function in terms of the partial molar Gibbs functions

$$G = \sum_i n_i \overline{G}_i = n\tilde{G} . \qquad (14.12)$$

We use the bar over the quantity and the subscript i to denote the partial molar quantity. We continue to use the tilde to represent a molar quantity and a circumflex to denote a quantity on a per-unit-mass basis.

In forming a partial molar quantity, we always start with an extensive quantity. Almost any extensive quantity works. For the heat capacities, we have

$$\overline{C}_{Vi} = \left(\frac{\partial C_V}{\partial n_i}\right)_{T,p,n_j \atop j\neq i} \text{ and } \overline{C}_{pi} = \left(\frac{\partial C_p}{\partial n_i}\right)_{T,p,n_j \atop j\neq i}. \qquad (14.13)$$

The result is then an intensive quantity.

The partial molar Gibbs function is given the special name, *chemical potential*, and the symbol μ_i:

$$\mu_i = \overline{G}_i = \left(\frac{\partial G}{\partial n_i}\right)_{T,p,n_j \atop j\neq i}. \qquad (14.14)$$

Let us express the differential of G as

$$dG = \left(\frac{\partial G}{\partial p}\right)_{T,n_j} dp + \left(\frac{\partial G}{\partial T}\right)_{p,n_j} dT + \sum_i \left(\frac{\partial G}{\partial n_i}\right)_{T,p,n_j \atop j\neq i} dn_i. \quad (14.15)$$

It follows from the definition 14.14 that the differential coefficients in the last term are the chemical potentials and from Eqs. 7.23 and 7.24 that those in the first two terms are the volume and minus the entropy. Hence, Eq. 14.15 can be written directly as

$$dG = Vdp - SdT + \sum_i \mu_i dn_i. \qquad (14.16)$$

In expressing the composition dependence of properties, the two most important characteristics of partial molar quantities are:

1. They permit extensive quantities to be expressed as in Eqs. 14.11 and 14.12.
2. Since the "logical" independent variables for the Gibbs function are the temperature and the pressure (see Table 7.1) and since these variables are held constant in the definition of the partial molar quantities, $\overline{G}_i$ is vested with a special significance. The differential of G can thus be put down immediately.

14.2 Mixing Experiments

We have merely defined the partial molar quantities; we have not considered how to measure them. To get at the composition dependence, we must perform mixing experiments, which are not required for the determination of k, β, and $\hat{C}_p$.

Figure 14.3 describes a mixing experiment where the system is isolated from the rest of the world; no heat is transferred and no work is performed. Consequently, the internal energy and the volume do not change during the experiment. By measuring the final temperature and pressure, one can obtain the internal energy U and volume V as functions of the temperature T, pressure p, and numbers n_{NH_3} and n_{H_2O} of moles of the components. We might notice that the numerical value of the internal energy of the mixture depends on the primary reference states chosen for both of the pure components.

For each final composition, only one mixing experiment needs to be performed. Subsequently, properties at other temperatures and pressures can be obtained by variations at constant composition, as described above.

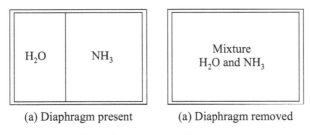

(a) Diaphragm present   (a) Diaphragm removed

Figure 14.3 Mixing with rigid, insulating walls.

If two materials, initially at the same temperature and pressure, are mixed and then brought back to the same temperature and pressure, the difference between the final and initial volumes is called the *volume change on mixing*. If the two materials are pure substances, such as ammonia and water, this volume change can be expressed as

$$\Delta V = n_{NH_3}(\overline{V}_{NH_3} - \widetilde{V}^{\circ}_{NH_3}) + n_{H_2O}(\overline{V}_{H_2O} - \widetilde{V}^{\circ}_{H_2O}), \qquad (14.17)$$

where the superscript $\circ$ denotes a pure component.

If the two materials, initially at the same temperature and pressure, are mixed so that the pressure remains constant and heat is removed or added so that the final temperature is the same as the initial value, the amount of heat added is called the *heat of mixing* (see Fig. 14.4). It follows from the analysis of constant-pressure processes in Chapter 4 that this is identical to the *enthalpy change on mixing*. If the two materials were pure, this result could be expressed as, for example,

$$Q = \Delta H = n_{NH_3}(\overline{H}_{NH_3} - \widetilde{H}^{\circ}_{NH_3}) + n_{H_2O}(\overline{H}_{H_2O} - \widetilde{H}^{\circ}_{H_2O}). \quad (14.18)$$

The mixing experiments described above could be irreversible without invalidating the results. An experiment for determining the composition dependence of the entropy is more difficult; the definition of the entropy requires one to measure the heat that would be transferred if the system were taken in a reversible manner from one state to another. This requires us to imagine that there exists a diaphragm or membrane that is permeable to only one component (see Fig. 14.5). This component is equilibrated across the membrane, while the other components are not.

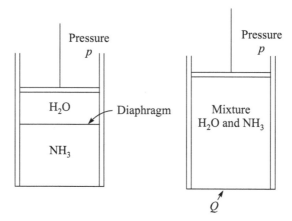

Figure 14.4 Mixing at constant pressure.

Suitable semipermeable membranes may not always present themselves. Experiments in phase equilibria are very useful in this regard. For example, Fig. 14.6 depicts equilibrium between a saline solution and pure water vapor. In this case, the phase boundary constitutes a semipermeable membrane since NaCl has a negligible tendency to enter the vapor phase. The composition of the liquid phase can be varied by condensation or evaporation, and the reversible heat can be measured.

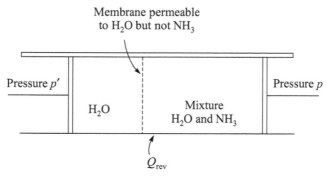

Figure 14.5 Experiment for reversibly adding a substance to or removing it from a mixture.

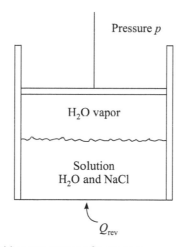

Figure 14.6 Reversible evaporation of a mixture.

Figure 7.1, from Example 7.2, shows an electrochemical system that permits reversible changes in the composition of the aqueous HCl solution. Consequently, this system can be used to obtain information about the entropy and enthalpy of these solutions as a function of composition. It also reveals information on chemical equilibria, since this system involves the reaction

$$\frac{1}{2}H_2 + AgCl \rightarrow HCl\,(aq) + Ag. \tag{14.19}$$

Problems

14.1 A partial molar quantity is an intensive property and depends only on intensive quantities, say the temperature, pressure, and composition, the last being expressed in terms of mole fractions. This is what allows the extensive properties to be expressed in terms of partial molar quantities as, for example, in Eqs. 14.11 and 14.12. Derive such a relationship using the entropy as an example.

   a. Write down the differential of the entropy, the independent variables being T, p, and n_i. Compare Eq. 14.15.

   b. Express n_i as $x_i n$ and integrate the differential at constant T, p, and x_i, making use of the fact that the partial molar entropies of the components are constant during this integration. (Thus, all the mole numbers are to vary at a rate proportional to their respective mole fractions.)

14.2 a. It is stated that it is sufficient to have the results of mixing experiments for the pressure (or equivalently a PVT equation of state), the enthalpy, and the entropy. Support this claim by showing how to evaluate the following properties on the basis of such mixing rules: (i) internal energy, (ii) Gibbs function, (iii) chemical potential, (iv) heat capacity at constant pressure.

   b. Why is it most difficult to obtain experimental values of the entropy of mixing, compared to the volume change on mixing and the enthalpy change on mixing?

   c. Why is the entropy change on mixing needed at only one temperature and pressure?

14.3 The specific enthalpy of mixtures of sulfuric acid and water at 25°C can be expressed approximately as

$$\hat{H} \text{ (Btu/lb)} = 44.5\, \omega_2 - 171.5\, \omega_1 \omega_2 - 687\, \omega_1 \omega_2^2,$$

where ω_1 is the mass fraction of sulfuric acid and ω_2 is the mass fraction of water.

   One pound of 30% acid is mixed isothermally with 1 pound of water. How much heat must be added to maintain the temperature at 25°C?

   If the heat capacity of 15% acid is 0.85 Btu/lb-°F, estimate the temperature change if the mixing process is adiabatic.

14.4 The specific enthalpy of mixtures of sulfuric acid and water at 25°C can be expressed approximately as

$$\widehat{H} \text{ (Btu/lb)} = 44.5\, \omega_2 - 171.5\, \omega_1\omega_2 - 687\, \omega_1\omega_2^2,$$

where ω_1 is the mass fraction of sulfuric acid and ω_2 is the mass fraction of water.

Determine expressions for the partial molar enthalpies of both sulfuric acid and water as functions of composition.

Notation

c_i	molar concentration of species i, mol/m^3
C_p	heat capacity at constant pressure, J/K
C_V	heat capacity at constant volume, J/K
G	Gibbs functions, J
H	enthalpy, J
k	isothermal compressibility, m^2/N
M_i	molar mass of species i, g/mol
n	total number of moles, mol
n_i	number of moles of species i, mol
p	pressure, N/m^2
Q	heat, J
S	entropy, J/K
T	thermodynamic temperature, K
U	internal energy, J
V	volume, m^3
x_i	mole fraction of species i
β	coefficient of thermal expansion, K^{-1}
μ_i	chemical potential of species i, J/mol
ρ	density, g/cm^3

Subscript, superscript, and special symbols

rev	reversible
°	pure component
~	per mole
^	per unit mass
—	partial molar

Reference

1. J. Willard Gibbs, "On the equilibrium of heterogeneous substances," *The Scientific Papers of J. Willard Gibbs* (New York: Dover Publications, Inc., 1961), Volume I, pp. 55–349. (Paper originally published from 1875 to 1878.)

Chapter 15

PVT Properties of Gas Mixtures

The accurate tabulation of properties for pure substances is bad enough because there are so many possible pure substances that can be formed from the elements. This measurement and tabulation of properties becomes even more impossible when we consider mixtures. Consequently, the prediction of properties of binary mixtures from data on pure substances and the prediction of properties of multicomponent mixtures on the basis of data on binary mixtures become empirical efforts with considerable economic incentive. In this chapter, we restrict ourselves to the PVT properties of gas mixtures.

15.1 Dalton's Law

Dalton's law says that pressures are additive when several gases are mixed in such a way that the initial volume of each gas before mixing is the same as the final volume of the mixture (see Fig. 15.1). The mixing is to be carried out at constant temperature:

$$p(T,V,n_1,n_2,...) = \sum_i p_i^\circ(T,V,n_i),\qquad(15.1)$$

where the superscript $\circ$ denotes the property of the pure substance under the conditions noted.

The Newman Lectures on Thermodynamics
John Newman and Vincent Battaglia
Copyright © 2019 Jenny Stanford Publishing Pte. Ltd.
ISBN 978-981-4774-26-0 (Hardcover), 978-1-315-10861-2 (eBook)
www.jennystanford.com

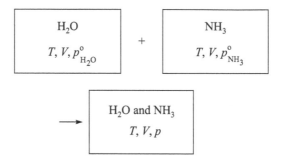

Figure 15.1 Pressures are additive according to Dalton's law.

15.2 Amagat's Law

Amagat's law says that the volumes are additive when several gases are mixed in such a way that the initial pressure of each gas before mixing is the same as the final pressure of the mixture (see Fig. 15.2). Again, the mixing is to be carried out at constant temperature:

$$V(T,p,n_1,n_2,...) = \sum_i V_i^\circ(T,p,n_i) = \sum_i n_i \tilde{V}_i(T,p). \quad (15.2)$$

| H$_2$O $T, p, V^\circ_{H_2O}$ | + | NH$_3$ $T, p, V^\circ_{NH_3}$ | → | H$_2$O and NH$_3$ T, p, V |

Figure 15.2 Volumes are additive according to Amagat's law.

This rule can also be phrased that the partial molar volumes of the components in the mixture are equal to the molar volumes of the pure components at the same temperature and pressure or that the volume change on mixing is zero (see Eq. 14.17).

Amagat's law and Dalton's law are approximate rules that allow the PVT properties of mixtures to be predicted on the basis of the properties of the pure components. Amagat's law would generally be expected to be more accurate than Dalton's law, although one might worry about applying the former when one of the components is a liquid at the temperature and pressure of the mixture.

The mixing at constant pressure according to Amagat's law (see Fig. 15.2) gives a better account of the effect of intermolecular forces since the intermolecular distances more nearly approximate those in the final mixture. For example, according to Dalton's law, one could mix 50 different gases, each at one atmosphere, to obtain a mixture at 50 atm. However, the conditions in the pure components at 1 atm are quite different from those in a system at 50 atm. Furthermore, one can convince oneself that Amagat's law is exact, while Dalton's law is only approximate, for mixing isotopes (in the absence of an isotope effect).

15.3   The Pseudo-critical Method

The corresponding-states correlation of compressibility factors (see Fig. 3.3) can be applied quite generally to gas mixtures. In forming the reduced pressure and temperature, one uses *pseudo-critical* properties, weighted averages of the properties for the pure components:[1]

$$p_{pc} = \sum_i y_i p_{ci} \text{ and } T_{pc} = \sum_i y_i T_{ci} , \qquad (15.3)$$

where p_{ci} is the critical pressure of component i (see Table 3.1), p_{pc} is the pseudo-critical pressure, etc.

15.4   Virial Equation

The PVT properties of a gas mixture can still be expressed by a virial equation (3.19 or 3.21). A rigorous result of statistical mechanics is that the second virial coefficient should be expressed as

$$B = \sum_i \sum_j y_i y_j B_{i,j} , \qquad (15.4)$$

where $B_{i,j}$ depends only on temperature and is, in fact, given by

$$B_{i,j} = B_{j,i} = -2\pi L \int_0^\infty [e^{-w_{i,j}/kT} - 1] r^2 dr , \qquad (15.5)$$

$w_{i,j}$ representing the intermolecular potential energy for a pair of molecules, one of species i and one of species j. (Compare Eq. 3.20

---

[1]In this and subsequent chapters, we shall try to follow chemical engineering practice by using y_i to denote the mole fraction in a vapor or gas phase and using x_i to denote the mole fraction in a liquid or in general.

and Fig. 3.4.) The values of $B_{i,i}$ are thus the same as the second virial coefficients of pure substances, treated in Chapter 3, and can be obtained from Fig. 3.5.

It is recommended that values of $B_{i,j}$ for pairs of different molecules can be obtained from Fig. 3.5 by taking the abscissa to represent $T/\sqrt{T_{ci}T_{cj}}$ and taking the ordinate to represent

$$8B_{i,j}\Big/(\tilde{V}_{ci}^{1/3} + \tilde{V}_{cj}^{1/3})^3 \;.$$

This is based on an extension of the microscopic principle of corresponding states, it being assumed that $\varepsilon_{i,j}$ and $\sigma_{i,j}$ (characterizing the intermolecular potential energy) can be approximated by

$$\varepsilon_{i,j} = \sqrt{\varepsilon_{i,i}\varepsilon_{j,j}} \quad \text{and} \quad \sigma_{i,j} = \frac{1}{2}(\sigma_{i,i} + \sigma_{j,j}). \tag{15.6}$$

We have discussed several *mixing rules* that permit the estimation of the density of a gas mixture on the basis of data on pure substances (or a corresponding-states correlation of data for pure substances). These methods are of great practical utility and also allow us to have definite prescriptions for classroom illustration, but they should be replaced by actual experimental data on the mixtures if this is possible and the most accurate results are sought.

Problems

15.1 Calculate the volume of a mixture of 1 mol of carbon dioxide and 1 mol of nitrous oxide (N_2O) at 148°F and 145 atm. Use three different methods—Dalton's law, Amagat's law, and the pseudo-critical method—and comment on the expected reliability of each result. You may express your answer in terms of the compressibility factor $Z_m = pV/nRT$ of the mixture.

15.2 Estimate the compressibility factor at 150°C and 343 atm of a mixture of 25% nitrogen and 75% hydrogen. State clearly what method you are using.

15.3 Suppose that Amagat's law is valid for gas-phase mixtures of components 1 and 2. What relationship does this imply for the second cross virial coefficient $B'_{1,2}$ in the pressure expansion?

Notation

B second virial coefficient, cm^3/mol

k Boltzmann constant, 1.38×10^{-23} J/K

L Avogadro's (or Loschmidt's) number, 6.0225×10^{23} mol^{-1}

n total number of moles, mol

n_i number of moles of species i, mol

p pressure, N/m^2

r intermolecular distance, cm

R universal gas constant, 8.3143 J/mol-K or 82.06 atm-cm^3/mol-K

T thermodynamic temperature, K

V volume, m^3

$w_{i,j}$ intermolecular potential energy, J

y_i mole fraction of species i

Z_m compressibility factor for mixture

$\varepsilon_{i,j}$ energy at minimum in intermolecular potential energy, J

$\sigma_{i,j}$ distance characteristic of intermolecular potential energy function, cm

Subscripts, superscript, and special symbol

c critical point

pc pseudo-critical property

$\circ$ pure component

$\sim$ per mole

Chapter 16

Thermodynamic Relationships

16.1 Maxwell Relations and Gibbs–Duhem Relation

In analogy with Chapter 7, we wish to consider some of the interrelationships among thermodynamic properties that arise on a formal mathematical basis.

Gibbs showed that an expression of the Gibbs function $G(T,p,n_j)$ as a function of temperature, pressure, and composition represents complete knowledge of the thermodynamic behavior of a multicomponent system. The entropy follows from Eq. 7.23, the volume from Eq. 7.24, and the chemical potentials from Eq. 14.14. The PVT equation of state is thus contained in the expression $G(T,p,n_j)$, and the heat capacity can be obtained by differentiating again the entropy with respect to temperature.

Furthermore, the expression $G(T,p,n_j)$ does not contain any redundant or inconsistent information. Possibilities for thermodynamic inconsistency have arisen already for pure substances. For example, it would be inconsistent to report that $\tilde{C}_p = 3.5R$ and $\tilde{C}_V = 2.9R$ for an ideal gas (see Eqs. 5.5 and 7.31). Likewise, it would be inconsistent to report a volume dependence of $\tilde{C}_V$ for a van der Waals fluid (see Example 7.4). In multicomponent

The Newman Lectures on Thermodynamics
John Newman and Vincent Battaglia
Copyright © 2019 Jenny Stanford Publishing Pte. Ltd.
ISBN 978-981-4774-26-0 (Hardcover), 978-1-315-10861-2 (eBook)
www.jennystanford.com

thermodynamics, we can avoid accusations of inconsistency by seeking the expression $G(T,p,n_j)$.

On the other hand, knowledge of entropy S, internal energy U, enthalpy H, or Helmholtz function A as a function of temperature, pressure, and composition does not constitute complete thermodynamic information. However, there are other complete sets. Suppose that entropy $S(U,V,n_j)$ is known as a function of internal energy, volume, and composition. Equation 7.25 yields p/T, and $1/T = (\partial S/\partial U)_V$. Now H, A, and G can be computed from the information available. Also, in statistical mechanics, thermodynamic properties are commonly inferred from an expression of the Helmholtz function $A(T,V,n_j)$ as a function of temperature, volume, and composition.

Let us look now at the Maxwell relations derivable from Eq. 14.16, the total differential of the Gibbs function. Equation 7.28 already relates the pressure dependence of the entropy to PVT properties. Similarly, the pressure dependence of the chemical potential is related to PVT properties since

$$\left(\frac{\partial \mu_i}{\partial p}\right)_{T,n_j} = \left(\frac{\partial V}{\partial n_i}\right)_{\substack{T,p,n_j \\ j \neq i}} = \bar{V}_i. \tag{16.1}$$

The temperature dependence of the chemical potential can also be obtained since

$$\left(\frac{\partial \mu_i}{\partial T}\right)_{p,n_j} = -\left(\frac{\partial S}{\partial n_i}\right)_{\substack{T,p,n_j \\ j \neq i}} = -\bar{S}_i. \tag{16.2}$$

(See Problem 16.1.) It is not immediately clear how to use this result since we do not know right away which is easier to measure, the temperature dependence of μ_i or the composition dependence of the entropy.[1] Differentiation again with respect to the temperature yields

$$\left(\frac{\partial^2 \mu_i}{\partial T^2}\right)_{p,n_j} = -\frac{\partial}{\partial n_i}\frac{\partial S}{\partial T} = -\frac{1}{T}\left(\frac{\partial C_p}{\partial n_i}\right)_{\substack{T,p,n_j \\ j \neq i}} = -\frac{1}{T}\bar{C}_{pi}. \tag{16.3}$$

Experimental measurements of the heat capacity as a function of composition and temperature can be helpful in the determination

---

[1]Note that the temperature derivative of the chemical potential has a limited direct physical significance since the value of $\bar{S}_i$ depends on the (somewhat arbitrary) primary reference state for the entropy of component i.

of chemical potentials at one temperature from those measured at another temperature.

One more Maxwell relation remains in Eq. 14.16. The second cross derivative of G with respect to mole numbers of two components yields

$$\left(\frac{\partial \mu_i}{\partial n_k}\right)_{T,p,n_j \atop j\neq k} = \frac{\partial^2 G}{\partial n_k \partial n_i} = \frac{\partial^2 G}{\partial n_i \partial n_k} = \left(\frac{\partial \mu_k}{\partial n_i}\right)_{T,p,n_j \atop j\neq i}. \qquad (16.4)$$

This expresses an important requirement of thermodynamic consistency. For example, in a binary system of components A and B, μ_A and μ_B are not independent functions of composition at a given temperature and pressure. One can be derived from the other, except for an additive constant.

Finally, let us form the differential of G from Eq. 14.12:

$$dG = \sum_i n_i d\mu_i + \sum_i \mu_i dn_i. \qquad (16.5)$$

Subtraction from the differential in Eq. 14.16 yields the so-called *Gibbs–Duhem relation*

$$SdT - Vdp + \sum_i n_i d\mu_i = 0. \qquad (16.6)$$

This equation has an important role in tests for thermodynamic consistency. It also expresses a relationship among intensive quantities and suggests the number of independent degrees of freedom for a system.

16.2 Gibbs Phase Rule and Phase Equilibria

In any problem, but particularly in a complicated problem with many variables, it is a good idea to enumerate explicitly the variables and the governing relations. The system is, in general, determinant only if the number of governing relations is the same as the number of variables. The system is not determined if there are fewer governing relations than variables, and the difference is the number of *degrees of freedom* that are still subject to specification. If there are fewer variables than governing relations, the system is indeterminant, and some new thinking is in order.

We are concerned mainly with intensive degrees of freedom. Extensive variables such as volume and numbers of moles deal with the size of the system and usually can be changed in obvious ways while keeping the intensive variables constant. For example, a pure substance has generally two (intensive) degrees of freedom. These can be thought of as temperature and pressure, although one could equally well specify $\hat{V}$ and $\hat{U}$. However, if we know that two phases are to be present, we lose a degree of freedom. At a given temperature, the pressure cannot be chosen at will but is determined by the vapor–pressure relation. The presence of three phases requires the substance to be at a triple point, with no choice in either temperature or pressure.

In these examples for a pure substance, any phase present could be increased in size without requiring a change in any of the intensive variables. Thus, the number of intensive and extensive degrees of remains three as we go from one phase to two phases to three phases.

Gibbs reflected on these matters for multicomponent systems and concluded that the number of intensive degrees of freedom F is given by

$$F = (C - R) - P + 2, \tag{16.7}$$

where C is the total number of components, R is the number of chemical reactions that can be equilibrated, and P is the number of phases present. Thus, $C - R$ is the number of *independent* components after taking into account possible chemical reactions. For example, O_2, CO, and CO_2 represent only two independent components if they can react according to

$$2\,CO + O_2 \rightleftharpoons 2\,CO_2. \tag{16.8}$$

If the temperature is so low that this reaction proceeds at a negligible rate, these can constitute three independent components. Conditions for chemical equilibria are taken up again in Chapter 19.

For a single phase, the degrees of freedom manifest themselves in our ability to specify the temperature, pressure, and composition. The last is expressed in the total number of moles and $C - 1$ intensive composition variables (see Eqs. 14.2 and 14.3). If these are expressed as mole fractions or mass fractions, it is apparent that there are only $C - 1$ independent values since the fractions must sum to unity. If these are expressed as molar concentrations $c_i = n_i/V$, it must be remembered that one of these is determined by the others and the temperature and pressure.

When two or more phases are present and equilibrated, their compositions are interrelated. Consider a component i that can pass freely from one phase α to another phase β. What is the equilibrium distribution of this component between the phases? For a closed system consisting of two or more phases but maintained at constant temperature and pressure, a spontaneous process must result in a reduction of the Gibbs function for the system. This conclusion was developed in Section 11.2 from the general condition that a spontaneous process results in an increase in the entropy of the universe.

At constant temperature and pressure, the change in the Gibbs function is

$$dG = dG^\alpha + dG^\beta = \sum_i \mu_i^\alpha dn_i^\alpha + \sum_i \mu_i^\beta dn_i^\beta, \qquad (16.9)$$

where the superscripts α and β designate the phases. Since there is a constant number of moles of component i in the system, $dn_i^\alpha = -dn_i^\beta$, and we can write

$$dG = \sum_i (\mu_i^\beta - \mu_i^\alpha) dn_i^\beta. \qquad (16.10)$$

As we continue to transfer component i from phase α to phase β, the Gibbs function G will continue to decrease as long as μ_i^β is less than μ_i^α, and the process will be spontaneous. When μ_i^β is greater than μ_i^α, there will be a tendency for component i to be transferred spontaneously in the opposite direction, and the condition of phase equilibrium is seen to be

$$\mu_i^\alpha = \mu_i^\beta \qquad (16.11)$$

for any component that can be simultaneously present in both phases. This condition is quite general and applies to all such components and applies also when more than two phases are involved.

A difference in the chemical potential μ_i is evidently the driving force for mass transfer from one phase to another. Indeed, it is also the driving force for diffusion within a phase. At equilibrium, the chemical potential will be uniform throughout any phase and will have the same value in each phase. The mole fractions must be uniform within a phase in order to ensure a constant value for each chemical potential.[2]

In a number of circumstances, it is convenient to select the chemical potentials μ_i as the intensive variables, in addition to the temperature and pressure. In view of Eq. 16.11, μ_i has a single value throughout an equilibrated system, whereas the mole fractions differ from one phase to another. Counting T, p, and μ_i, we have $C + 2$ intensive variables. The Gibbs–Duhem relation 16.6 provides a constraint among these variables, since it can be written in the intensive form

$$\tilde{S}dt - \tilde{V}dp + \sum_i x_i d\mu_i = 0. \tag{16.12}$$

Because the composition is different in each phase, this equation can be written independently for each phase, thus providing a total of P constraints on the intensive variables. Remaining is the same number of degrees of freedom as expressed in the Gibbs phase rule 16.7.

Problems

16.1 Show that

$$\left(\frac{\partial(\mu_i/T)}{\partial T}\right)_{p,n_j} = -\frac{\overline{H}_i}{T^2}$$

and that

$$\mu_i = \overline{H}_i - T\overline{S}_i.$$

16.2 Suppose that you just derived an expression for the Helmholtz function A in terms of the temperature, volume, and numbers of moles of components. Describe how you would now compute values for any desired thermodynamic property, including the heat capacities, the chemical potentials, the isothermal compressibility k, and the coefficient of thermal expansion β.

16.3 The partial molar volumes in a binary liquid are reported to be fit accurately by the expressions

---

[2]These conclusions are modified for a phase subjected to a gravitational or centrifugal field, where composition differences will exist even at equilibrium.

$$\overline{V}_1 - \tilde{V}_1^{\circ} = ax_2^2 + bx_2^2(4x_1 - 1),$$

$$\overline{V}_2 - \tilde{V}_2^{\circ} = ax_1^2 - bx_1^2(3 - 4x_1).$$

Are these expressions thermodynamically consistent? Show the basis for your answer.

Notation

A	Helmholtz free energy, J
c_i	molar concentration of species i, mol/m^3
C	number of components
C_p	heat capacity at constant pressure, J/K
F	number of intensive degrees of freedom
G	Gibbs function, J
H	enthalpy, J
k	isothermal compressibility, m^2/N
n_i	number of moles of species i, mol
p	pressure, N/m^2
P	number of phases
R	number of independent, equilibrated chemical reactions
S	entropy, J/K
T	thermodynamic temperature, K
U	internal energy, J
V	volume, m^3
x_i	mole fraction of species i
β	coefficient of thermal expansion, K^{-1}
μ_i	chemical potential of species i, J/mol

Superscripts and special symbols

α, β	phases α and β
$\sim$	per mole
$\wedge$	per unit mass
—	partial molar

Chapter 17

Ideal-Gas Mixtures

At low pressures, the ideal-gas law 3.8

$$pV - nRT = RT \sum_i n_i \qquad (17.1)$$

is obeyed by mixtures as well as pure substances. The readers can verify that this is in harmony with both Dalton's law 15.1 and Amagat's law 15.2.

The *Gibbs mixing rule* tells us the essential results of mixing experiments with ideal gases: a property of the mixture is the sum of the properties of the pure components, where these latter properties are evaluated under conditions that the same number n_i of moles of species i that are in the mixture occupy the same volume V and have the same temperature as the mixture (see Fig. 17.1).

For the pressure, we write

$$p = \sum_i p_i = \sum_i p_i(T,V,n_i). \qquad (17.2)$$

This is identical to Dalton's law 15.1 of additive pressures and the ideal-gas law 17.1 and really tells us nothing new. The term *partial pressure* can be used to denote p_i, the pressure exerted by species i if it were alone in the same volume as the mixture. From the ideal-gas law, the partial pressure is

$$p_i = n_i RT/V = pn_i/n = py_i, \qquad (17.3)$$

The Newman Lectures on Thermodynamics
John Newman and Vincent Battaglia
Copyright © 2019 Jenny Stanford Publishing Pte. Ltd.
ISBN 978-981-4774-26-0 (Hardcover), 978-1-315-10861-2 (eBook)
www.jennystanford.com

where y_i is the mole fraction of component i. The term partial pressure is sometimes used in a different sense for nonideal systems. For the enthalpy, we have

$$H = \sum_i H_i^\circ (T,V,n_i) = \sum_i n_i \tilde{H}_i^\circ (T).$$ (17.4)

$p_1 = y_1 p$		$p_2 = y_2 p$
T, V, n_1	$+$	T, V, n_2

$p = $ total pressure
T, V, n_1, n_2

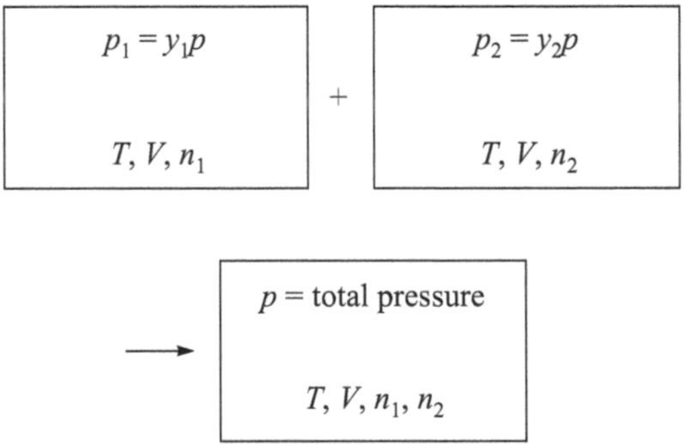

Figure 17.1 According to the Gibbs mixing rule, the entropy of a mixture of ideal gases is equal to the sum of the entropies of the pure components if each pure component individually occupied the same volume as the mixture and was maintained at the same temperature. A similar additive property holds for the pressure and the enthalpy. From these mixing rules, any other thermodynamic property of the mixture can be derived and calculated from thermodynamic properties of the pure components.

This result is moderately simple since the molar enthalpy $\tilde{H}_i^\circ$ of an ideal gas depends only on the temperature, and it is related to the heat capacity by

$$\tilde{H}_i^\circ (T) = \tilde{H}_i^* (T) = \int^T \tilde{C}_{pi}^* \, dT .$$ (17.5)

The lower limit of integration is arbitrary and depends on the reference state in which the enthalpy of pure component i is taken to be zero. This reference state can be different for different components. The asterisk is used to designate quantities that are related to the ideal-gas state. In this case $\tilde{C}_{pi}^*$ is the zero-pressure limit of the molar heat capacity (at constant pressure) of component i and was treated in Chapter 5.

These two examples of the Gibbs mixing rule, applied to the pressure and the enthalpy, have a simple physical meaning; they say that there is no volume change on mixing and no heat of mixing for ideal gases (see Eqs. 14.17 and 14.18). They also tell us that the partial molar enthalpy of a component in the mixture is equal to the molar enthalpy of that component at the same temperature,

$$\overline{H}_i = \left(\frac{\partial H}{\partial n_i}\right)_{T,p,n_j \atop j\neq i} = \widetilde{H}_i^*(T), \tag{17.6}$$

and that the partial molar volume is equal to the molar volume of that component at the temperature and pressure of the mixture,

$$\overline{V}_i = \widetilde{V}_i^*(T,p) = RT/p. \tag{17.7}$$

These results are readily understandable in terms of molecular concepts. The molecules of an ideal gas behave rather independently of each other, and this situation continues to prevail after they are mixed, if the final pressure is still low enough. Thus, it makes sense that on mixing at constant volume (as depicted in Fig. 17.1), the pressures should be additive and the internal energies should be additive. It then follows that the enthalpies would be additive:

$$H = U + pV = \sum_i n_i \widetilde{U}_i^*(T) + V p_i(T,V,n_i)$$

$$= \sum_i n_i \widetilde{U}_i^*(T) + n_i RT = \sum_i n_i \widetilde{H}_i^*(T). \tag{17.8}$$

The situation is not quite so simple for the entropy. The Gibbs mixing rule yields

$$S = \sum_i S_i^o(T,V,n_i) = \sum_i n_i \widetilde{S}_i^*(T,p_i). \tag{17.9}$$

Note that it is at the partial pressure p_i that the molar entropy of component i must be evaluated in order for the entropies to be additive as indicated.

In view of the fundamental equations (7.28 and 7.19)

$$\left(\frac{\partial S}{\partial p}\right)_T = -\left(\frac{\partial V}{\partial T}\right)_p \text{ and} \left(\frac{\partial S}{\partial T}\right)_p = \frac{C_p}{T}, \tag{17.10}$$

the molar entropy of a substance i, which behaves like an ideal gas, can be expressed as

$$\widetilde{S}_i^*(T,p) = -R \ln p + \int^T \frac{\widetilde{C}_{pi}^*}{T} dT, \tag{17.11}$$

and Eq. 17.9 becomes

$$S = \sum_i n_i \left[-R \ln p_i + \int^T \frac{\tilde{C}_{pi}^*}{T} dT \right]$$

$$= \sum_i n_i \left[-R \ln p y_i + \int^T \frac{\tilde{C}_{pi}^*}{T} dT \right]$$

$$= \sum_i n_i \tilde{S}_i^*(T,p) - \sum_i n_i R \ln y_i. \qquad (17.12)$$

The partial molar entropy of component *i* thus is

$$\bar{S}_i = \tilde{S}_i^*(T,p) - R \ln y_i, \qquad (17.13)$$

and the *entropy of mixing* is not zero:

$$\Delta S = S - \sum_i n_i \tilde{S}_i^*(T,p) = -\sum_i n_i R \ln y_i. \qquad (17.14)$$

A plausible demonstration of the validity of the Gibbs mixing rule for entropy can be advanced. Figure 17.2 shows two interconnected boxes, which can slide into one another without friction. At the ends of the boxes are membranes permeable to one or the other of the components. This is a reversible process for effecting the isothermal mixing depicted in Fig. 17.1. If the Gibbs mixing rule applies to the pressures, then the pressures add as indicated on Fig. 17.2. This means that the net force on each box is zero, and hence the work W is zero for the process. For example, the force on the right box is

$$p_A^{\mathcal{A}} + p_B^{\mathcal{A}} - (p_A + p_B)\mathcal{A} = 0, \qquad (17.15)$$

where $\mathcal{A}$ is the area of an end of the box.

The heat Q supplied to the system is to be calculated from the energy balance:

$$\Delta U = Q - W. \qquad (17.16)$$

If the Gibbs mixing rule applies to the internal energy, then ΔU is zero for this isothermal process. Since ΔU and W are both zero, it follows that Q is zero and thus also ΔS:

$$\Delta S = Q/T = (\Delta U + W)/T = 0. \qquad (17.17)$$

This result is just the Gibbs mixing rule for the entropy, and thus the demonstration is complete.

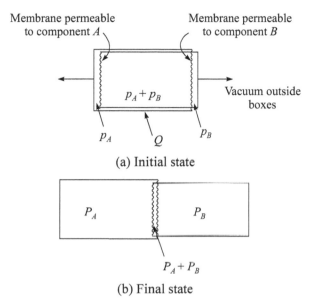

(a) Initial state

(b) Final state

Figure 17.2 Process for the reversible, isothermal separation of a two-component mixture into its components. (Compare Fig. 14.5.)

(A likely candidate for a semipermeable membrane for two gaseous components may not readily suggest itself. Figure 17.3 shows how an electrochemical cell might act as a membrane permeable to oxygen but not to nitrogen. In order to regard the process as reversible, one would have to ignore the diffusion of nitrogen across the system, the ohmic potential drop, the concentration changes of KOH due to the fact that potassium ions carry about 25% of the current, and the overpotential at the oxygen electrodes, which are not noted for being reversible.)

Mixing rules for the pressure, enthalpy, and entropy, as given above, should be sufficient, since mixing rules for other quantities such as the heat capacity, the internal energy, and the Gibbs function can be obtained from those for p, H, and S.

For example, for the heat capacity we have

$$C_p = \left(\frac{\partial H}{\partial T}\right)_{p,n_i} = \sum_i n_i \tilde{C}_{pi}^*(T) = \sum_i C_{pi}^o(T,V,n_i). \qquad (17.18)$$

Thus, we see that the Gibbs mixing rule also applies to heat capacity. For the Gibbs function, we have

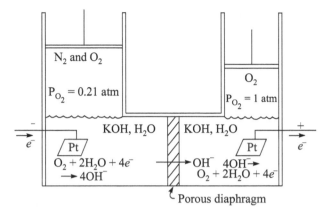

Figure 17.3 Semipermeable membrane for oxygen, constructed from an electrochemical cell.

$$G = H - TS = \sum_i n_i \left[\int^T \tilde{C}_{pi}^* dT - T \int^T \frac{\tilde{C}_{pi}^*}{T} dT + RT \ln p y_i \right]$$

$$= \sum_i n_i \tilde{G}_i^\circ (T,p) + RT \sum_i n_i \ln y_i \qquad (17.19)$$

where

$$\tilde{G}_i^\circ (T,p) = \tilde{H}_i^* (T) - T \tilde{S}_i^* (T,p) = RT \ln p - \int^T \int^T \frac{\tilde{C}_{pi}^*}{T} dT dT . \qquad (17.20)$$

The *Gibbs free energy of mixing*, therefore, is

$$\Delta G = G - \sum_i n_i \tilde{G}_i^\circ (T,p) = RT \sum_i n_i \ln y_i , \qquad (17.21)$$

and the chemical potential of component k is

$$\mu_k = \left(\frac{\partial G}{\partial n_k} \right)_{\substack{T,p,n_j \\ j \neq k}} = \tilde{G}_k^\circ (T,p) + RT \ln y_k$$

$$= \mu_k^* (T) + RT \ln p_k , \qquad (17.22)$$

where

$$\mu_k^* (T) = - \int^T \int^T \frac{\tilde{C}_{pk}^*}{T} dT dT \qquad (17.23)$$

is the temperature-dependent part of the chemical potential of pure component k (in the ideal-gas state). The numerical value of $\mu_k^*(T)$ depends on the primary reference states chosen for both the enthalpy and the entropy of the pure component k. (See also Example 17.4.)

Notice that the chemical potential does not follow the Gibbs mixing rule since

$$\mu_k \neq \sum_i \mu_{k,i}^0(T,V,n_i), \qquad (17.24)$$

$\mu_{k,i}^0$ being equal to minus infinity unless $i = k$. The Gibbs mixing rule appears to apply to extensive quantities and to the pressure but not, in general, to intensive quantities.

Example 17.1 Ideal gases are mixed irreversibly by removing the separating diaphragm (see Fig. 17.4). The heat capacity of both nitrogen and oxygen can be taken to be constant and given by $\tilde{C}_p^* = 7R/2$. Calculate the entropy change of the system when the gases on the two sides are allowed to mix, by removal of the diaphragm.

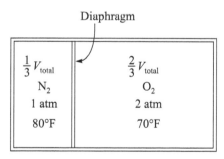

Diaphragm

$\frac{1}{3}V_{total}$	$\frac{2}{3}V_{total}$
N_2	O_2
1 atm	2 atm
80°F	70°F

Figure 17.4 Irreversible mixing of ideal gases. The outer walls are rigid and insulating.

Solution: As in Example 7.5, we take 1 mol of the final mixture as a basis, and we let the initial left side be denoted by 1 and the initial right side by 2. The mole numbers and the final temperature and pressure of the mixture are still those calculated in Example 5.3 since there is no heat of mixing for ideal gases.

Equation 17.12 gives the entropy of an ideal-gas mixture. The entropy change for the system can thus be expressed as

$$\Delta S = n_1 \left[\tilde{C}_p^* \ln \frac{T_f}{T_1} - R \ln \frac{p_f n_1}{p_1 n} \right] + n_2 \left[\tilde{C}_p^* \ln \frac{T_f}{T_2} - R \ln \frac{p_f n_2}{p_2 n} \right].$$

This differs from the result calculated in Example 7.5 only by an *entropy of mixing* term. Thus,

$$\frac{\Delta S}{n} = 0.0911 - R \frac{n_1}{n} \ln \frac{n_1}{n} - R \frac{n_2}{n} \ln \frac{n_2}{n}$$

$$= 0.0911 + 0.9846 = 1.076 \text{ Btu/lb-mol-°R.}$$

We see here that the entropy of mixing is 10 times larger than the entropy change that can be associated with the pressure and temperature changes.

Example 17.2 Reversible work to remove nitrogen from air. We have available a large supply of air, for which we may take the mole fraction of oxygen to be 0.21 and that for nitrogen to be 0.79. This is to be passed through a separator, taken to be reversible, in order to produce purified oxygen. This purified oxygen is subsequently burned with propane in a power plant, the nitrogen content being reduced in order to reduce the amount of nitrogen oxides released as pollutants to the atmosphere. We wish to estimate the increase in the cost of power (see Example 17.3), but first let us focus our attention on the separator (see Fig. 17.5) by calculating the reversible work required per mole of product oxygen.

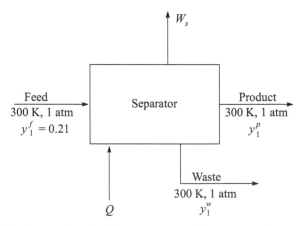

Figure 17.5 Separation of air into an oxygen-rich stream and an oxygen-lean stream.

Solution: Let component 1 be oxygen and component 2 be nitrogen. We take as our basis, in essence, 1 mol of product oxygen, and we wish to calculate $-W_s/n_1^p RT$. There are two parameters subject to later specification. One is the purity y_1^p of the product, oxygen-rich stream, and the other is the size of the waste stream that we are willing to reject back to the atmosphere. This latter is represented by n^w/n^p, the ratio of the total numbers of moles in the waste and product streams.

We proceed with balances of entropy, energy, and material. For a reversible process,[1]

$$Q = T\Delta S,$$

and for a steady-flow process

$$\Delta H = Q - W_s.$$

Hence, the reversible work for this steady-flow process is related to the change in the Gibbs function

$$W_s = T\Delta S - \Delta H = -\Delta G.$$

Substitution of Eq. 17.19 yields

$$-\frac{W_s}{RT} = n_1^p \ln\frac{y_1^p}{y_1^f} + n_1^w \ln\frac{y_1^w}{y_1^f} + n_2^p \ln\frac{y_2^p}{y_2^f} + n_2^w \ln\frac{y_2^w}{y_2^f},$$

where we have made use of the material balances

$$n_1^f = n_1^p + n_1^w \text{ and } n_2^f = n_2^p + n_2^w$$

and the fact that the temperatures and pressures are the same for the feed, product, and waste streams. (Incidentally, for this process, ΔH is itself equal to zero, and $Q = W_s$.)

With the material balances and the known value of y_1^f, we are now able to construct Table 17.1, which gives $-W_s/n_1^p RT$ for given values of the parameters y_1^p and n^w/n^p.

High product purity is toward the top in Table 17.1, and this involves more work. The second column, for n^w much larger than n^p, corresponds to a large waste stream whose composition is essentially identical to that of the feed. This involves the least work for a given product purity. The entry for n^w/n^p equal to 3.76 corresponds to a

---

[1]The temperature here should really be the temperature of the reversible heat source, which we shall take to be 300 K.

waste stream of pure nitrogen and a product stream of pure oxygen. The entries for y_1^p less than 0.21 show that work is required even if the product stream is depleted in oxygen.

Table 17.1 $-W_s/n_1^p RT$, the reversible work for separating oxygen from nitrogen, as a function of y_1^p and n^w/n^p, the product mole fraction and the ratio of waste stream to product stream.

	$-W_s/n_1^p RT$		
y_1^p	$n^w \gg n^p$	$n^w = 5n^p$	$n^w = 3.76n^p$
1.0	1.56	2.07	2.46
0.99	1.506	1.997	
0.9	1.226	1.627	
0.7	0.7890	1.028	
0.5	0.4104	0.5196	
0.4	0.2317	0.2888	
0.1	0.4313	0.5025	
0.05	2.069	2.367	

One might speculate about the ratio of the actual work of separation to that indicated by the reversible work in Table 17.1. This ratio, which is always greater than 1, depends on the particular nature of the irreversibilities involved in the process. Here the process will involve liquefaction and distillation of air, and there will be lost work associated with compression, refrigeration, and heat exchange (see Example 12.3). High, but not perfect, product purity can be achieved by including more stages in the distillation, and this will contribute to the capital costs. Furthermore, the cost becomes very large when $n^w \gg n^p$ simply because a large quantity of air must then be handled. Optimum values of y_1^p and n^w/n^p are dictated on this basis of cost coupled with the expected benefit to be achieved for various degrees of product purity.

For these reasons, thermodynamic analysis alone cannot give a complete picture of the feasibility of a proposed process. However, it can provide minimum values for the work required and should thus be used for preliminary screening of new schemes. If the reversible work is unattractive, no further effort is needed to eliminate a scheme from further consideration.

Actual processes for air separation require work about seven times that of the reversible work. This ratio could, of course, be reduced further but with an increase in the capital costs.

Example 17.3 Balances on a power plant. Use the results of Example 17.2 to estimate the increase in the cost of electric power if purified oxygen is used rather than air in the combustion process. Assume that propane is burned with the stoichiometric amount of the product stream from the gas separator (see Fig. 17.6), that the heating value of propane is 2.1 MJ/mol, and that 40% of this energy can be converted to electrical power, W_p, at a cost of 4.0 cent/kWh.

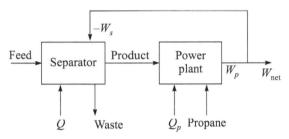

Figure 17.6 An electric power generating plant consumes the oxygen-rich stream from the separator of Example 17.2, at the same time supplying the separator with the power required to effect the separation.

Solution: The combustion reaction is

$$C_3H_8 + 5O_2 \rightarrow 3CO_2 + 4H_2O.$$

Under the stated conditions, 1 mol of oxygen thus produces 0.168 MJ of electric power or

$$W_p = 1.68 \times 10^5 \, n_1^p = W_{net} - W_s = W_{net} + n_1^p RT(-W_s / n_1^p RT).$$

Thus,

$$\frac{W_{net}}{W_p} = 1 - \frac{RT}{1.68 \times 10^5} \left(\frac{-W_s}{n_1^p RT} \right) = 1 - 0.015 \left(\frac{-W_s}{n_1^p RT} \right).$$

The cost of power is increased in the ratio

$$\frac{W_p}{W_{net}} = \frac{1}{1 - 0.015(-W_s / n_1^p RT)}.$$

Thus, for $-W_s / n_1^p RT = 1.56$, the cost of power is increased by a factor of 1.024 from 4.0 to 4.1 cent/kWh, an increase of 2.4%. If the actual work for the separator is seven times the reversible work, then the cost of power must be increased by 19%, even with no allowance for the capital cost of the separator.

If the denominator of this expression should go to zero, the separator would be consuming all the power produced in the power plant, and none would remain to be sold.

Example 17.4 Develop an explicit expression for $\mu_i^*(T)$ based on the temperature dependence of the ideal-gas heat capacity.

Solution: Let the ideal-gas heat capacity be written as

$$\tilde{C}_{pi}^*(T) = a_i + b_i T + c_i / T^2 + \gamma_i T^2.$$

This is a generalization of Eq. 5.6 to include a quadratic term in T, often used as an alternative to the term c_i / T^2 in fitting the function. Integration gives the molar enthalpy of an ideal gas:

$$\tilde{H}_i^*(T) = \tilde{H}_i^*(T_o) + a_i(T - T_o) + \frac{b_i}{2}(T^2 - T_o^2)$$

$$- c_i\left(\frac{1}{T} - \frac{1}{T_o}\right) + \frac{\gamma_i}{3}(T^3 - T_o^3),$$

where the value of the enthalpy at a reference temperature T_o is involved in the integration constant.

According to the relation

$$\left(\frac{\partial(\mu_i / T)}{\partial T}\right)_{p,n_j} = -\frac{\overline{H}_i}{T^2}$$

(see Problem 16.1), division by T^2 and a second integration yield μ_i^*

$$\frac{\mu_i^*}{T} = \frac{\mu_i^*(T_o)}{T_o} + \tilde{H}_i^*(T_o)\left(\frac{1}{T} - \frac{1}{T_o}\right) - a_i\left(\ln\frac{T}{T_o} + \frac{T_o}{T} - 1\right)$$

$$- \frac{b_i}{2}\left(T - T_o + \frac{T}{T_o^2} - T_o\right)$$

$$- \frac{c_i}{2}\left(\frac{1}{T^2} - \frac{1}{T_o^2} - \frac{2}{T_o T} + \frac{2}{T_o^2}\right) - \frac{\gamma_i}{6}\left(T^2 - T_o^2 + \frac{2T_o^3}{T} - 2T_o^2\right).$$

Rearrangement allows this to be written in the simpler form

$$\mu_i^*(T) = \frac{T}{T_0}\mu_i^*(T_0) + \left(1 - \frac{T}{T_0}\right)\tilde{H}_i^*(T_0) - a_i\left(T\ln\frac{T}{T_0} + T_0 - T\right)$$
$$-\frac{(T-T_0)^2}{2}\left(b_i + \frac{c_i}{TT_0^2} + \frac{T+2T_0}{3}\gamma_i\right).$$

This result shows clearly the presence of two integration constants, $\mu_i^*(T_0)$ and $\tilde{H}_i^*(T_0)$, dependent ultimately on the primary reference states for enthalpy and entropy. For values of T close to T_0, the terms in a_i, b_i, c_i, and γ_i become very small (quadratic in $T - T_0$).

Problems

17.1 Two ideal gases, both at the same pressure p, are in an insulated container and separated by an insulated partition. Calculate the change in pressure, internal energy, and entropy and the final temperature when the partition is removed. Gas 1, of n_1 moles, has a constant heat capacity $\tilde{C}_{p1}^*$ and an initial temperature T_1, and gas 2, of n_2 moles, has a different constant heat capacity $\tilde{C}_{p2}^*$ and an initial temperature T_2.

17.2 Integrate the heat capacity directly according to Eq. 17.23 to obtain for $\mu_i^*(T)$ the same expression derived in Example 17.4. This requires explicit identification of the integration constants.

17.3 a. With the ideal-gas heat capacity expressed as in Example 17.4, integrate according to Eq. 17.11 to obtain the entropy $\tilde{S}_i^*(T,p)$.

b. Combine the entropy from Problem (a) with the molar enthalpy derived in Example 17.4 to obtain the chemical potential μ_i.

c. Integrate the entropy from Problem (a) according to Eq. 16.2 to obtain the chemical potential. This should agree with the result obtained in Problem (b).

d. Now, from Problem (b) or (c), identify $\mu_i^*(T)$ according to its definition in Eq. 17.22. This result should be identical with the expression for $\mu_i^*(T)$ derived in Example 17.4.

17.4   Prove the following equalities of integrals of the heat capacity:

$$\int_{T_o}^{T} \tilde{C}_{pi}^{*}(T')dT' - T\int_{T_o}^{T} \frac{\tilde{C}_{pi}^{*}(T')}{T'}dT'$$

$$= -\int_{T_o}^{T}\int_{T_o}^{T'} \frac{\tilde{C}_{pi}^{*}(T'')}{T''}dT''\,dT' = -T\int_{T_o}^{T} \frac{1}{(T')^2}\int_{T_o}^{T'}\tilde{C}_{pi}^{*}(T'')\,dT''\,dT'.$$

Notation

$\mathcal{A}$ area, m^2
C_p heat capacity at constant pressure, J/K
G Gibbs function, J
H enthalpy, J
n total number of moles, mol
n_i number of moles of species i, mol
p pressure, N/m^2
p_i partial pressure of species i, N/m^2
Q heat, J
R universal gas constant, 8.3143 J/mol-K or 82.06 atm-cm^3/mol-K
S entropy, J/K
T thermodynamic temperature, K
U internal energy, J
V volume, m^3
W work, J
W_s shaft work, J
y_i mole fraction of species i
μ_i chemical potential of species i, J/mol
μ_i^{*} secondary reference state quantity for ideal-gas state, J/mol

Subscripts, superscripts, and special symbols

f feed
i component i
p product
w waste
$\circ$ pure component
$*$ ideal-gas state
$\sim$ per mole
$\wedge$ per unit mass
$-$ partial molar

Chapter 18

Fugacity Coefficient

In Chapter 17, the thermodynamic properties of ideal-gas mixtures were developed. This is a convenient place to develop the properties of real gases because departures from the ideal-gas equation of state (Eq. 17.1) can be used directly to obtain the required correction factors by integration of Eq. 7.24 and related equations. This situation is discussed in Section 14.2, where the importance of mixing experiments is emphasized, in Chapter 17, where it is pointed out that the Gibbs mixing rule provides the results of such experiments for ideal gases, and in Section 16.1, where we observe that thermodynamic properties of mixtures of a given composition could be calculated at different pressures and temperatures by integration of the partial molar volume or the partial molar heat capacity. Furthermore, PVT properties of gases are introduced briefly in Chapter 15.

An analogous situation exists for pure components. *Departure functions* for enthalpy and entropy were introduced in Chapter 12. Figures 12.4 and 12.5 present these results for substances that obey a corresponding-states equation of state. For multicomponent systems, there is an increased emphasis on the Gibbs function and on the chemical potentials both for phase equilibria (see Eq. 16.11) and for chemical equilibria (see Eq. 19.5). Consequently, in this

The Newman Lectures on Thermodynamics
John Newman and Vincent Battaglia
Copyright © 2019 Jenny Stanford Publishing Pte. Ltd.
ISBN 978-981-4774-26-0 (Hardcover), 978-1-315-10861-2 (eBook)
www.jennystanford.com

chapter, we present the fugacity coefficient ϕ_i (defined by Eq. 18.5), a departure function for the chemical potential. We do this first for the virial equation of state, which is particularly appropriate for gases at moderate pressures, and next for Amagat's law. Last, we present fugacity coefficients for a corresponding-states substance over a range of reduced temperatures and pressures.

18.1 The Virial Equation

Integration of Eq. 7.24

$$\left(\frac{\partial G}{\partial p}\right)_{T,n_i} = V \tag{18.1}$$

for a gas mixture obeying the virial Eq. 3.21 yields

$$\tilde{G} = F(T,n_i) + RT\left(\ln p + B'p + \frac{1}{2}C'p^2 + \frac{1}{3}D'p^3 \cdots\right), \tag{18.2}$$

where the integration constant F can depend on temperature and composition. Comparison at low pressures with Eq. 17.19 shows that

$$F(T,n_i) = \sum_i y_i[\mu_i^*(T) + RT \ln y_i], \tag{18.3}$$

where $\mu_i^*(T)$ is an integral of the ideal-gas limit of the heat capacity of pure component i according to Eq. 17.23.

In using the results of Chapter 17 to evaluate $F(T, n_i)$, we are, in effect, making use of mixing experiments in the ideal-gas state, which are described by the Gibbs mixing rule. Equation 18.1 then describes variations with pressure after the mixing has been carried out. Equations 18.2 and 18.3 thus yield G for all temperatures, pressures, and compositions at which the virial equation adequately describes the PVT data. A knowledge of G as a function of T, p, and n_i allows one to calculate all the thermodynamic properties of the mixture.

From Eqs. 18.2 and 18.3, the chemical potential μ_i is calculated according to Eq. 14.14 to be

$$\mu_i = \mu_i^*(T) + RT \ln y_i p$$

$$+pRT \left(\frac{\partial}{\partial n_i} \right)_{T,p,n_j \atop j \neq i} \left(nB' + \frac{n}{2}C'p + \frac{n}{3}D'p^2 + \cdots \right). \tag{18.4}$$

It is customary to define the *fugacity coefficient* ϕ_i so that, in general,

$$\mu_i = \mu_i^*(T) + RT \ln(y_i p \phi_i). \tag{18.5}$$

In other words, the fugacity coefficient describes departures from the ideal-gas state, and ϕ_i approaches 1 in low-pressure mixtures. With the virial equation, the fugacity coefficient thus becomes

$$\ln \phi_i = p \left(\frac{\partial}{\partial n_i} \right)_{T,p,n_j \atop j \neq i} \left(nB' + \frac{n}{2}C'p + \frac{n}{3}D'p^2 + \cdots \right). \tag{18.6}$$

The second virial coefficient is given rigorously by (see Eq. 15.4)

$$B' = \sum_i \sum_j y_i y_j B'_{i,j}. \tag{18.7}$$

Substitution into Eq. 18.6, with neglect of terms involving C', D', etc., yields

$$\ln \phi_i = p \sum_j 2 y_j B'_{i,j} - B'p, \tag{18.8}$$

and for a two-component mixture, this becomes

$$\ln \phi_1 = p[y_1(2 - y_1)B'_{1,1} + 2y_2^2 B'_{1,2} - y_2^2 B'_{2,2}]. \tag{18.9}$$

These equations should be accurate up to about 10 atm, and the method for estimating the B' values is outlined in Chapters 3 and 15.

18.2 Amagat's Law

Amagat's law involves mixing at constant temperature and pressure and says that the volume change on mixing is zero or

$$V = \sum_i n_i \tilde{V}_i^\circ(T,p). \tag{18.10}$$

Consequently, the partial molar volume is

$$\bar{V}_i = \tilde{V}_i^\circ(T,p) \tag{18.11}$$

and is the same as the molar volume of pure component i at the same temperature and pressure.

Equation 16.1 shows that the partial molar volume holds the key to the pressure dependence of the chemical potential:

$$\left(\frac{\partial \mu_i}{\partial p}\right)_{T,n_j} = \overline{V}_i. \tag{18.12}$$

Integration for a mixture obeying Amagat's law yields

$$\mu_i = \int^p \tilde{V}_i^{\circ} dp + f_i(T,n_j), \tag{18.13}$$

where it is recognized that the integration constant can depend on temperature and composition, the variables held constant in the differentiation in Eq. 18.12. However, with the assumption of Amagat's law, the integral in Eq. 18.13 is independent of composition and must, therefore, be related to the chemical potential of pure component i at the given temperature and pressure. Evaluation of Eq. 18.13 for pure component i yields

$$\mu_i^{\circ} = \int^p \tilde{V}_i^{\circ} dp + F_i(T), \tag{18.14}$$

where F_i is the value of f_i evaluated for pure component i.

Now the chemical potential of pure component i can be expressed from Eq. 18.5 as

$$\mu_i^{\circ} = \mu_i^{*}(T) + RT \ln (p\phi_i^{\circ}), \tag{18.15}$$

where ϕ_i° is the fugacity coefficient of pure component i at the given temperature and pressure. Substitution for the integral in Eq. 18.13 gives

$$\mu_i = \mu_i^{\circ} - F_i(T) + f_i(T,n_j)$$

$$= \mu_i^{*}(T) + RT \ln (p\phi_i^{\circ}) - F_i(T) + f_i(T,n_j). \tag{18.16}$$

Recall from Eq. 17.22 that, in the limit of zero pressure, the chemical potential takes the form

$$\mu_i \rightarrow \mu_i^{*}(T) + RT \ln (py_i). \tag{18.17}$$

Since ϕ_i° also approaches unity in this limit, it becomes possible to evaluate the integration constant:

$$f_i(T, n_j) - F_i(T) = RT \ln y_i. \tag{18.18}$$

The chemical potential now reads

$$\mu_i = \mu_i^*(T) + RT \ln(py_i\phi_i^o). \tag{18.19}$$

Comparison with Eq. 18.5 shows that if a mixture obeys Amagat's law, the fugacity coefficient is equal to the value for pure component *i* at the same temperature and pressure:

$$\phi_i = \phi_i^o(T, p). \tag{18.20}$$

This result is called the *Lewis fugacity rule.*

18.3 Corresponding-States Correlation

The fugacity coefficient for a pure component is a function describing a departure from the ideal-gas behavior and can be calculated solely from the PVT equation of state; it is independent of the ideal-gas heat capacities. Consequently, it can be approximated by means of a corresponding-states correlation in terms of reduced temperature and pressure (reduced with the critical properties). Such a graph, shown in Fig. 18.1, is particularly useful when Amagat's law and the Lewis fugacity rule can be applied to a mixture. The graph in Fig. 18.1 is directly related to the departure functions for enthalpy and entropy (shown in Figs. 12.4 and 12.5), and all three are related to the corresponding-states correlation of the compressibility factor (see Fig. 3.3).

18.4 Other Equations of State

Thermodynamic properties of nonideal gas mixtures can be developed from other equations of state. Chueh and Prausnitz [1] followed, in some detail, the consequences of the Redlich–Kwong Eq. 3.13, including the best "mixing rules" by which to estimate the Redlich–Kwong parameters *a* and *b* for the mixture from values for pure components as well as any experimental information that might be available on binary mixtures.

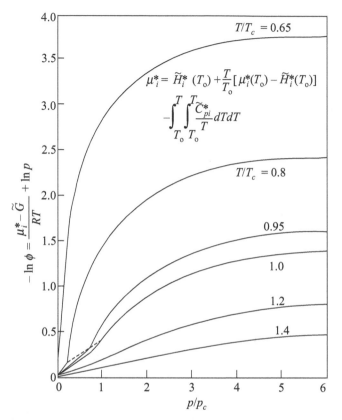

Figure 18.1 Fugacity coefficient for pure fluids. While the data are for nitrogen (see Figs. 12.4 and 12.5), they are plotted in a form made dimensionless with the critical properties so that they can be applied to any corresponding-states fluid. There is a two-phase line, not a region, in this graph; it is a dashed line that should extend from the origin to about $\ln \phi = -0.393$ at the critical point.

Problems

18.1 Integrate Eq. 7.24 in the form

$$\left(\frac{\partial G}{\partial V}\right)_{T,n_i} = V\left(\frac{\partial p}{\partial V}\right)_{T,n_i}$$

for a gas mixture obeying the virial Eq. 3.19. Use this result to show that the fugacity coefficient ϕ_i can be expressed as

$$\ln \phi_i = -\ln\left(1 + \frac{B}{\widetilde{V}} + \frac{C}{\widetilde{V}^2} + \frac{D}{\widetilde{V}^3} + \cdots\right) + \frac{B}{\widetilde{V}} + \frac{C}{\widetilde{V}^2} + \frac{D}{\widetilde{V}^3} + \cdots + \frac{1}{\widetilde{V}}\left(\frac{\partial nB}{\partial n_i}\right)_{T,n_j \atop j\neq i}$$

$$+ \frac{1}{2\widetilde{V}^2}\left(\frac{\partial nC}{\partial n_i}\right)_{T,n_j \atop j\neq i} + \frac{1}{3\widetilde{V}^3}\left(\frac{\partial nD}{\partial n_i}\right)_{T,n_j \atop j\neq i} + \cdots .$$

18.2 Show that $\mu_i^*(T)$ can be expressed as

$$\mu_i^*(T) = K_1 + K_2(T - T_o) - \int_{T_o}^{T}\int_{T_o}^{T}\frac{\widetilde{C}_{pi}^*}{T}dTdT \,,$$

where K_1 and K_2 are constants and the limits of integration are now given by some definite temperature T_o.

Let the primary reference state for water be specified by the requirement that both the enthalpy and the entropy be zero for saturated liquid water at the triple point, 0.01°C. Letting $T_o = 0.01°C = 273.16$ K, shows unambiguously how to evaluate the constants in the above expression for $\mu_i^*(T)$ for water. Bear in mind that Eq. 18.2 involves taking the logarithm of the pressure p, which is not dimensionless.

18.3 Since the mole fractions sum to 1, all the mole fractions cannot realistically be used as independent variables. In solution theory, one therefore identifies one component as the *solvent* and uses as independent, intensive variables the temperature, the pressure, and the mole fractions of the other components, now known as *solutes*.

For a three-component gas mixture, let component 1 be the solvent. With neglect of virial coefficients beyond the second, show that the fugacity coefficient of a solute species can be expressed as, for example,

$$\ln \phi_2 = \ln \phi_2^\theta + 2p(\beta'_{2,2}y_2 + \beta'_{2,3}y_3)$$

$$-p(\beta'_{2,2}y_2^2 + 2\beta'_{2,3}y_2y_3 + \beta'_{3,3}y_3^2)$$

and that of the solvent is

$$\ln \phi_1 = \ln \phi_1^o - p(\beta'_{2,2}y_2^2 + 2\beta'_{2,3}y_2y_3 + \beta'_{3,3}y_3^2) \,,$$

where

$$\beta'_{i,j} = B'_{i,j} + B'_{1,1} - B'_{1,i} - B'_{1,j}$$

is, in a sense, a virial coefficient relative to the solvent. Here ϕ_2^θ represents the fugacity coefficient of species 2 in nearly pure solvent ($y_1 \to 1$) and is given by

$$\ln \phi_2^\theta = p(2B'_{1,2} - B'_{1,1}),$$

while ϕ_1^o is the fugacity coefficient of the pure solvent, given by

$$\ln \phi_1^o = pB'_{1,1}.$$

Notice that for dilute solutions, ϕ_1 departs much less from ϕ_1^o than ϕ_2 does from ϕ_2^θ. Furthermore, the Lewis fugacity rule is much better for the solvent than for a solute since $\phi_2^\theta \neq \phi_2^o$.

In developing approximate methods for dilute solutions, one would be tempted to drop the last terms in $\ln\phi_2$, those which are quadratic in the small quantities y_2 and y_3. If this were done here and in the corresponding form for $\ln \phi_3$, would the resulting formulation be thermodynamically consistent?

18.4 For a dilute solution of components 2 and 3 in a solvent of component 1 (as in Problem 18.3), let us use instead mole ratios $r_i = n_i/n_1 = y_i/y_1$ for the solute species as the independent composition variables. Express the chemical potential of a solute, like component 2, as

$$\mu_2 = \mu_2^\theta + RT \ln r_2 + RT \ln (y_1\phi_2 / \phi_2^\theta),$$

where

$$\mu_2^\theta(T,p) = \mu_2^*(T) + RT \ln (p\phi_2^\theta)$$

is a secondary reference state quantity for an infinite-dilution reference state and $RT\ln(y_1\phi_2 / \phi_2^\theta)$ represents a correction term, supposed to be small in dilute solutions. (ϕ_2^θ is given in Problem 18.3.)

Show that the correction term can be expressed as

$$\ln (y_1\phi_2 / \phi_2^\theta) = 2(\beta_{2,2}r_2 + \beta_{2,3}r_3)$$

if virial coefficients beyond the second are ignored and if quadratic and higher order terms in the small quantities r_2 and r_3 are ignored. Here

$$\beta_{i,j} = p(B'_{i,j} + B'_{1,1} - B'_{1,i} - B'_{1,j}) - 0.5.$$

Show that this truncated form for $\ln(y_1\phi_i/\phi_i^\theta)$ for solute species is thermodynamically consistent and that the corresponding form for the chemical potential of the solvent is

$$\mu_1 = \mu_1^o + RT\ln(y_1\phi_1/\phi_1^o)$$

where the correction term is

$$\ln(y_1\phi_1/\phi_1^o) = -\sum_{i=2}^{C} r_i - \sum_{i=2}^{C}\sum_{j=2}^{C}\beta_{i,j}r_ir_j.$$

(ϕ_1^o was given in Problem 18.3.)

What is the corresponding form for the total Gibbs function G of the mixture?

18.5 For a pure fluid, show that

$$-\ln\phi = \frac{\widetilde{H}^* - \widetilde{H}}{RT_c}\frac{T_c}{T} - \frac{\widetilde{S}^* - \widetilde{S}}{R}$$

and how it should, therefore, be possible, in principle, to obtain ϕ for Fig. 18.1 from the data in Figs. 12.4 and 12.5.

For a pure fluid, show that

$$\frac{\widetilde{S}^* - \widetilde{S}}{R} = \left(\frac{\partial(T\ln\phi)}{\partial T}\right)_p,$$

$$\frac{\widetilde{H}^* - \widetilde{H}}{RT_c} = \frac{T^2}{T_c}\left(\frac{\partial\ln\phi}{\partial T}\right)_p,$$

and

$$Z = 1 + p\left(\frac{\partial\ln\phi}{\partial p}\right)_T$$

and how it should, therefore, be possible, in principle, to obtain Figs. 3.3, 12.4, and 12.5 all from the data in Fig. 18.1.

Do these results mean that it is possible to obtain $\ln\phi$ from Fig. 12.4 alone?

Notice that the equation given above for Z, coupled with the definitions inherent in the virial equation itself, implies that

$$B' = \frac{B}{RT} = \lim_{p\to 0}\left(\frac{\partial\ln\phi}{\partial p}\right)_T.$$

Thus the initial slopes in Fig. 18.1 should be the same as those on Fig. 3.3, as suggested below Eq. 3.25.

18.6 Obtain an expression for the fugacity coefficient of a pure substance that obeys the equation of state:

$$p = \frac{RT}{\tilde{V} - b}.$$

18.7 Obtain an expression for the fugacity coefficient of a pure substance that obeys the van der Waals equation of state. The thermodynamic properties of such a substance were developed in Example 7.4.

18.8 Obtain an expression for the fugacity coefficient of a pure substance that obeys the Redlich–Kwong equation of state. (See also Problem 7.4.)

18.9 Estimate the fugacity coefficients for the following pure substances at 400°C and 250 atm:

a. H_2 d. CH_3OH

b. CO   e. CH_4

c. CO_2 f. H_2O

18.10 Estimate the fugacity coefficients for the following pure substances at 150°C and 350 atm:

a. H_2 c. NH_3

b. N_2

18.11 Calculate the fugacity, that is, the partial pressure times the fugacity coefficient, for steam at 800°F and 2000 psia. You are expected to use the steam tables. If you use a different method, be sure to state what it is.

18.12 Show how to determine the temperature derivative of the logarithm of the fugacity coefficient. Would you need any heat-capacity data in order to evaluate this quantity?

18.13 A mixture of A and B is equilibrated with pure A through a semipermeable membrane, as sketched in Fig. 18.2. What does the semipermeable membrane do by way of relating properties in the compartment of pure A to those of the mixture? If the gas mixture obeys the truncated virial equation, indicate how you would go about determining the pressure of pure A on the left side of the membrane. The composition of the mixture is specified.

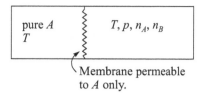

Figure 18.2 Equilibration through a semipermeable membrane.

18.14 Estimate the error in using the Lewis fugacity rule for calculating the fugacity coefficient of CO_2 at 39 atm and 25°C when the mole fraction of H_2O is 0.05. (Actually the water probably condenses under these conditions.) The second virial coefficient for CO_2 is -125 cm^3/mol and that for water is -1300 cm^3/mol. You need to estimate the second cross virial coefficient.

Notation

B second virial coefficient, cm^3/mol

B' second virial coefficient in pressure expansion, atm^{-1}

$B'_{i,j}$ pair-wise contribution to B', atm^{-1}

C number of components

C, D third and fourth virial coefficients

C', D' third and fourth virial coefficients in pressure expansion

C_p heat capacity at constant pressure, J/K

f_i, F integration constants

G Gibbs function, J

H enthalpy, J

n total number of moles, mol

n_i number of moles of species i, mol

p pressure, atm

r_i mole ratio $= n_i/n_1$

R universal gas constant, 8.3143 J/mol-K or 82.06 atm-cm^3/mol-K

S entropy, J/K

T thermodynamic temperature, K

V volume, m^3

y_i mole fraction of species i

Z compressibility factor $= p\tilde{V}/RT$

$\beta_{i,j}$ coefficient in expansion of correction terms

$\beta'_{i,j}$ coefficient in expansion of correction terms, atm^{-1}

μ_i chemical potential of species i, J/mol

μ_i^* secondary reference state quantity for ideal-gas state, J/mol

ϕ_i fugacity coefficient of species i

Subscripts, superscripts, and special symbols

c critical point

i species i

$\circ$ pure component

θ dilute solution in a specified solvent

$*$ ideal-gas state

$\sim$ per mole

—   partial molar

Reference

1. P. L. Chueh and J. M. Prausnitz, "Vapor-liquid equilibria at high pressures. Vapor-phase fugacity coefficients in nonpolar and quantum-gas mixtures," *Industrial and Engineering Chemistry Fundamentals*, **6**, 492–498 (1967).

Chapter 19

Gas-Phase Reactions

Chemical conversions are governed by both thermodynamics and kinetics, or rate processes. We concern ourselves here with the general conditions of chemical equilibrium and the method of calculation of the equilibrium compositions on the basis of a few tabulated thermodynamic data.

19.1 Condition for Equilibrium

The general condition for equilibrium is that any spontaneous process must result in an increase in the entropy of the universe. However, since physical-property data are frequently given in terms of temperature and pressure, a more direct route may be to imagine the system to be in contact with a reversible heat source at the temperature T and in contact with a reversible work source so that the boundaries of the system are maintained at a constant pressure p. Then, as shown in Section 11.2, spontaneous processes will lead to a minimization of the Gibbs function G for the system.

Consider two possible chemical reactions, represented abstractly as

$$0 \rightarrow \sum_i \nu_{i1} M_i \quad \text{and} \quad 0 \rightarrow \sum_i \nu_{i2} M_i. \qquad (19.1)$$

The Newman Lectures on Thermodynamics
John Newman and Vincent Battaglia
Copyright © 2019 Jenny Stanford Publishing Pte. Ltd.
ISBN 978-981-4774-26-0 (Hardcover), 978-1-315-10861-2 (eBook)
www.jennystanford.com

Here M_i stands for the chemical formula of species i, and v_{i1} is the stoichiometric coefficient of species i in reaction 1. For example, for the ammonia synthesis

$$N_2 + 3H_2 \rightarrow 2NH_3, \tag{19.2}$$

v_i is -1 for nitrogen, -3 for hydrogen, and 2 for ammonia. The stoichiometric coefficient will be zero for nonreactants and catalysts. The chemical reaction equation can be multiplied through by a constant without affecting the equilibrium composition. Hence, it is ratios of stoichiometric coefficients that are important.

The differential of the Gibbs function of the system is (see Eqs. 14.15 and 14.16)

$$dG = \left(\frac{\partial G}{\partial T}\right)_{p,n_j} dT + \left(\frac{\partial G}{\partial p}\right)_{T,n_j} dp + \sum_i \left(\frac{\partial G}{\partial n_i}\right)_{T,p,n_j \atop j \neq i} dn_i$$

$$= -SdT + Vdp + \sum_i \mu_i dn_i. \tag{19.3}$$

Some constraints are placed upon the system. The temperature and pressure have been fixed. Furthermore, in a closed system, the mole numbers can change only in harmony with the stoichiometry of the reactions. Let the *extent of reaction* 1 be denoted by ξ_1 and that of reaction 2 be denoted by ξ_2, such that $v_{i1}d\xi_1/dt$ is the rate of production of species i in reaction 1. Thus, for these two reactions at constant temperature and pressure, Eq. 19.3 becomes

$$dG = \left(\sum_i v_{i1}\mu_i\right)d\xi_1 + \left(\sum_i v_{i2}\mu_i\right)d\xi_2. \tag{19.4}$$

We see that with these constraints, there are only two independent degrees of freedom, one associated with the extent to which each reaction occurs. Minimization of G implies that dG is zero for arbitrary variations of ξ_1 and ξ_2 near the equilibrium composition. This will be true only if the coefficients of $d\xi_1$ and $d\xi_2$ are zero. Thus, the condition for chemical equilibrium is that

$$\sum_i v_{i1}\mu_i = 0 \tag{19.5}$$

and similarly for reaction 2. One can perceive that the derivation would also be applicable for any other possible reactions. Gibbs also emphasized the point that once the system is equilibrated, it is immaterial whether that condition was approached at constant temperature and pressure or, for example, at constant volume and internal energy. Thus, equation is quite a general condition for chemical equilibrium and applies to a liquid mixture as well as a gas, and it is true even when a heterogeneous catalytic reaction surface may be present.

19.2   Equilibrium Composition

Practical application requires the determination of the composition from Eq. 19.5. Substitution of Eq. 18.5 for the chemical potentials in a gas mixture yields (for more than one reaction, an additional subscript can be added to v_i and K):

$$\sum_i v_i \mu_i^*(T) + \sum_i v_i RT \ln (y_i p\phi_i) = 0 \tag{19.6}$$

or

$$\prod_i (y_i p\phi_i)^{v_i} = K \tag{19.7}$$

where

$$\ln K = -\sum_i v_i \frac{\mu_i^*(T)}{RT} = -\frac{\Delta G^*}{RT}. \tag{19.8}$$

K is called the *equilibrium constant,* and $\Delta G^* = \sum_i v_i \mu_i^*$ is the standard Gibbs-function change for the reaction as written in Eq. 19.1 or 19.2. If Eq. 19.2 were divided by 2, so as to produce 1 mol of ammonia, the stoichiometric coefficients would change and ΔG^* would have half as large a numerical value. The pressure cancels in Eq. 19.7 only if the number of molecules produced in the reaction is the same as the number of those that are consumed. Otherwise, K has the dimension of pressure raised to a power determined by the stoichiometric coefficients. The unit of pressure must be the same as that used in the determination of the numerical value of μ_i^*. The *atmosphere* is the commonly used pressure unit.

K is truly an equilibrium constant at a given temperature. It contains the dominant temperature dependence of the equilibrium, as discussed in the next section. Auxiliary equilibrium ratios are often employed. K_p is the ratio of partial pressures $p_i = py_i$,

$$K_p = \prod_i (y_i p)^{v_i} = K \prod_i (\phi_i)^{-v_i} , \qquad (19.9)$$

and reduces to K when departures of the fugacity coefficients from unity can be ignored. The principal pressure and composition dependence is retained, since ϕ_i is not a strong function of these variables. Inclusion of the fugacity coefficients can be regarded as a refinement of the calculations that might be desirable at very high pressures. K_y is the ratio of mole fractions,

$$K_y = \prod_i (y_i)^{v_i} = K \prod_i (p\phi_i)^{-v_i}, \qquad (19.10)$$

and may be useful in calculations at a given total pressure.

For example, for the ammonia synthesis reaction 19.2, the equilibrium condition is

$$\mu_{N_2} + 3\mu_{H_2} = 2\mu_{NH_3}, \qquad (19.11)$$

but this is dealt with more conveniently in the form

$$K = \frac{1}{p^2} \frac{y_{NH_3}^2}{y_{N_2} y_{H_2}^3} \frac{\phi_{NH_3}^2}{\phi_{N_2} \phi_{H_2}^3} = \exp\left(-\frac{\Delta G^*}{RT}\right), \qquad (19.12)$$

where

$$\Delta G^* = 2\mu_{NH_3}^* - \mu_{N_2}^* - 3\mu_{H_2}^* . \qquad (19.13)$$

Now, K has units of atm^{-2}, which is related to the fact that μ_i^* involves the logarithm of pressure, a dimensional quantity.[1]

Let us turn now to the calculation of the equilibrium composition. Suppose that K has been determined and the pressure p has been selected for ammonia synthesis. With neglect of fugacity coefficients, we can write

$$K_y = \frac{y_{NH_3}^2}{y_{N_2} y_{H_2}^3} = p^2 K . \qquad (19.14)$$

Let the initial mole fractions of nitrogen and hydrogen be $y_{N_2}^0$ and $y_{H_2}^0$. Then the numbers of moles change according to the stoichiometry as follows:

---

[1]Conceptually, it is wise to define λ_i^* by $\mu_i^* = RT \ln \lambda_i^*$ so that $\mu_i = RT \ln (\lambda_i^* p y_i \phi_i)$. The units of λ_i^* are then reciprocal to those of p. However, μ_i^* is usually found tabulated, not λ_i^*.

$$dn_{N_2} = -n^o d\xi, \quad dn_{H_2} = -3n^o d\xi,$$

$$dn_{NH_3} = 2n^o d\xi, \quad dn = -2n^o d\xi, \tag{19.15}$$

where ξ is an intensive extent of reaction made dimensionless with the initial total number of moles n^o. The mole fractions can be expressed as

$$y_{H_2} = \frac{n^o y_{H_2}^o - 3n^o \xi}{n^o - 2n^o \xi} = \frac{y_{H_2}^o - 3\xi}{1 - 2\xi},$$

$$y_{H_2} = \frac{y_{N_2}^o - \xi}{1 - 2\xi}, \quad y_{NH_3} = \frac{y_{NH_3}^o + 2\xi}{1 - 2\xi}. \tag{19.16}$$

Substitution into Eq. 19.14 gives a single equation for the determination of the extent of reaction:

$$K_y = \frac{\left(y_{NH_3}^o + 2\xi\right)^2 (1 - 2\xi)^2}{\left(y_{N_2}^o - \xi\right)\left(y_{H_2}^o - 3\xi\right)^3}. \tag{19.17}$$

It is becoming apparent that these calculations can be complicated, especially if several reactions are possible, and a computer solution may be appropriate. We return to this point in Example 19.3. In the meantime, consider the interesting case of a stoichiometric mixture of nitrogen and hydrogen, where $y_{N_2}^o = 0.25$ and $y_{H_2}^o = 0.75$. The extent of reaction ξ can then range between 0 and 0.25, and we define the *conversion* C here as $C = 4\xi$. For the stoichiometric mixture, Eq. 19.17 can be solved to yield the simple result

$$C = 1 - \frac{2}{\sqrt{4 + 3\sqrt{3K_y}}}. \tag{19.18}$$

Conversion is plotted against K_y in Fig. 19.1. K_y varies over about five orders of magnitude as conversion varies from 10% to 90%, and complete conversion is achieved only with an infinite value of K_y. The equilibrium ratio K_y itself is proportional to p^2 and also depends strongly on temperature. The temperature and pressure of the reactor can thus be chosen with the limitations of the equilibrium conversion in mind. This will be treated in more detail in Example 19.2.

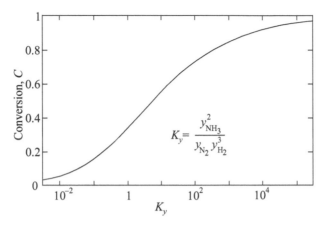

Figure 19.1 Equilibrium conversion to ammonia in a stoichiometric nitrogen–hydrogen mixture, as a function of the equilibrium ratio K_y.

19.3 Equilibrium Constant

Equations 19.12 and 19.13 define the equilibrium constant K in terms of the functions $\mu_i^*(T)$. These functions were defined by Eq. 17.23 and can be written as

$$\mu_i^*(T) = \int^T \tilde{C}_{pi}^* dT - T \int^T \frac{\tilde{C}_{pi}^*}{T} dT .\qquad (19.19)$$

Two integration constants still need to be specified. These correspond to the primary reference state for the enthalpy and entropy, as developed in the case of a single material in Problem 18.2. In chemical thermodynamics, the primary reference states cannot be chosen arbitrarily for each material; values used for components entering into a chemical reaction must be in harmony. This should be clear from Eqs. 19.12 and 19.13, since the equilibrium constant itself cannot depend on arbitrary reference states.

Table 19.1 gives values of μ_i^* for a number of materials at 25°C. In practice, primary reference states are chosen for the elements. The first three rows in Table 19.1 indicate this choice for hydrogen, nitrogen, and oxygen. Elemental carbon cannot be vaporized; its primary reference state is indicated in Table 19.2. The stable form of sulfur at 25°C and 1 atm, rather than the ideal-gas state, is chosen for its primary reference state. Tables of chemical thermodynamic properties, from which our tables are abstracted, collect values based on many experimental measurements over the years. Entries may be given for several states for one material. Table 19.1 covers the ideal-gas state; Table 19.2 includes some of the same materials in their stable state at 1 atm, and data for dilute solutions in given solvents can be found in the tables.

Since the temperature derivative of the Gibbs function is $-S$, we can write

$$\left(\frac{\partial(G/T)}{\partial T}\right)_{p,n_j} = -\frac{G}{T^2} - \frac{S}{T} = -\frac{H}{T^2}. \qquad (19.20)$$

Similarly, we can obtain from Eq. 19.19

$$\frac{d(\mu_i^*/T)}{dT} = -\frac{1}{T^2}\int^T \tilde{C}_{pi}^* dT = -\frac{\tilde{H}_i^*(T)}{T^2}. \qquad (19.21)$$

Since values for the enthalpy are given in Tables 19.1 and 19.2, we have a second integration constant for μ_i^*. Primary reference states are again specified for the elements. Now μ_i^* and $\tilde{H}_i^*$ can be determined at temperatures other than 25°C by appropriate integration of the heat capacity $\tilde{C}_{pi}^*$, for which extensive tables exist (see, for example, Table 3.1 as well as Ref. [3]). The enthalpy is, of course, important in its own right for many energy calculations such as the heat released on combustion and the adiabatic flame temperature. The integration for μ_i^* is carried out in Example 17.4 for a fairly general expression for the heat capacity.

Equation 19.21 also provides the route to the temperature derivative of the equilibrium constant:

$$\frac{d\ln K}{dT} = -\frac{d\Delta G^*/RT}{dT} = \frac{\Delta H^*}{RT^2} \qquad (19.22)$$

where

$$\Delta H^* = \sum_i v_i \tilde{H}_i^*(T) \qquad (19.23)$$

is the enthalpy change for the reaction at constant temperature and a low pressure. An exothermic reaction is one for which ΔH^* is negative. For such a reaction, K decreases as T increases, according to Eq. 19.22. This decreases the equilibrium conversion, in harmony with the principle of Le Châtelier.

Table 19.1 Enthalpies and standard Gibbs function in the ideal-gas state (1 atm pressure unit for μ_i^*) at 25°C, with primary reference states based on the elements.

	$\widetilde{H}_i^*$ kcal/mol	μ_i^* kcal/mol	$\widetilde{C}_{pi}^*$ cal/mol-K
hydrogen (H_2)	0	0	6.889
nitrogen (N_2)	0	0	6.961
oxygen (O_2)	0	0	7.016
sulfur (S_2)	30.68	18.96	7.76
methane (CH_4)	−17.88	−12.13	8.439
carbon dioxide (CO_2)	−94.051	−94.254	8.87
acetylene (HCCH)	54.19	50.00	10.50
hydrazine (NH_2NH_2)	22.80	38.07	11.85
carbon monoxide (CO)	−26.416	−32.780	6.959
nitric oxide (NO)	21.57	20.69	7.133
nitrogen dioxide (NO_2)	7.93	12.26	8.89
nitrous oxide (N_2O)	19.61	24.90	9.19
ammonia (NH_3)	−11.02	−3.94	8.38
water (H_2O)	−57.796	−54.634	8.025
hydrogen sulfide (H_2S)	−4.93	−8.02	8.18
sulfur dioxide (SO_2)	−70.944	−71.748	9.53
sulfur trioxide (SO_3)	−94.58	−88.69	12.11
carbon disulfide (CS_2)	28.05	16.05	10.85
ethane (C_2H_6)	−20.24	−7.86	12.58
ethylene (C_2H_4)	12.49	16.28	10.41
methanol (CH_3OH)	−47.96	−38.72	10.49
acetonitrile (CH_3CN)	20.9	25.0	12.48
methyl mercaptan (CH_3SH)	−5.34	−2.23	12.01
nitromethane (CH_3NO_2)	−17.86	−1.65	13.70
methyl amine (CH_3NH_2)	−5.49	7.67	12.7
acetic acid (CH_3COOH)	−103.31	−89.4	15.9

ethanol (C₂H₅OH)	−56.19	−40.29	15.64
ethylene oxide (C₂H₄O)	−12.58	−3.12	11.45
cyanogen (NCCN)	73.84	71.07	13.58
nitric acid (HNO₃)	−32.28	−17.87	12.75
hydrogen cyanide (HCN)	32.3	29.8	8.57
formaldehyde (HCHO)	−28.0	−27.0	8.46

Note: Molar heat capacities are also given. Data are taken from Ref. [1]. Remember that 1 cal = 4.184 J.

Table 19.2 Enthalpy, Gibbs function, and molar heat capacity at 25°C and 1 atm, with primary reference states based on the elements.

	$\widetilde{H}^{\circ}$ kcal/mol	μ_i° kcal/mol	$\widetilde{C}_p$ cal/mol-K
C graphite	0	0	2.038
S rhombic	0	0	5.41
N₂H₄ liquid	12.10	35.67	23.63
H₂O liquid	−68.315	−56.687	17.995
SO₂ liquid	−76.6		
SO₃ I, β crystal	−108.63	−88.19	
SO₃ liquid	−105.41	−88.04	
CS₂ liquid	21.44	15.60	18.1
CH₃OH liquid	−57.04	−39.76	19.5
CH₃CN liquid	12.8	23.7	21.86
CH₃SH liquid	−11.08	−1.85	21.64
CH₃NO₂ liquid	−27.03	−3.47	25.33
CH₃NH₂ liquid	−11.3	8.5	
CH₃COOH liquid	−115.8	−93.2	29.7
C₂H₅OH liquid	−66.37	−41.80	26.64
C₂H₄O liquid	−18.60	−2.83	21.02
HNO₃ liquid	−41.61	−19.31	26.26
HCN liquid	26.02	29.86	16.88

Note: Data taken from Ref. [1].

A number of equilibrium situations can be summarized concisely by means of Tables 19.1 and 19.2. For example, ΔG^*, in Eq. 19.13, for the ammonia synthesis is twice the numerical value given for ammonia in Table 19.1. At 25°C, we calculate $K = 5.98 \times 10^5$ atm^{-2}. By rewriting Eq. 19.22 in the equivalent form

$$\frac{\partial \ln K}{\partial (1/T)} = -\frac{\Delta H^*}{R}, \qquad (19.24)$$

we are inspired to plot the logarithm of K versus the reciprocal of temperature in Fig. 19.2. The enthalpy change of the reaction, $\Delta H^* = -22.04$ kcal/mol (again from Table 19.1), provides the slope of the curve at 25°C. The curve itself is very nearly a straight line. It is similar to the vapor pressure graph (Fig. 11.3). Thus, the whole curve can be reproduced reasonably well from just a few values in Table 19.1.

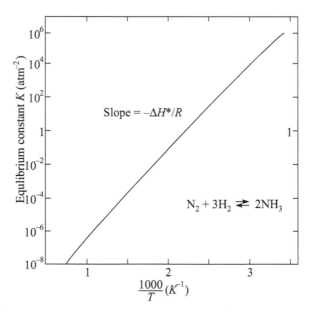

Figure 19.2 Temperature dependence of the equilibrium constant for the ammonia synthesis.

The curve in Fig. 19.2 of the logarithm of K versus the reciprocal of the temperature is not a perfect straight line because the enthalpy change ΔH^* of the reaction is not strictly constant. In other words, the heat capacity of the products is different from the heat capacity of the reactants. Table 19.1 allows us to calculate, at 25°C,

$$\Delta C_p^* = \sum_i v_i \tilde{C}_{pi}^* = -10.87 \text{ cal/mol-K}. \qquad (19.25)$$

To assess the magnitude of this effect, we might note that the tangent line through the point at 25°C would predict that $K = 10.09$ atm^{-2} at 150°C, while the actual value lies below this by 24%. On the other hand, a temperature change of only 4.6°C will change K by the same factor.

19.4 Independent Reactions

For complicated situations where many species are present, it is important to have an orderly procedure for identifying the correct number of independent reactions, a set of such reactions, and a set of independent material balances, necessary for the quantitative determination of the equilibrium composition. One procedure is as follows:

1. Write down a chemical reaction for the formation from the elements of each species present. Be sure to include reactions such as

$$3O_2 \rightleftharpoons 2O_3$$

and

$$H_2 \rightleftharpoons 2H$$

if both forms of an element are possibly present.

2. Choose an element that appears in the reaction equations but is not considered to be a species actually present in the system. Use one of the reaction equations to eliminate this elemental species from the remaining reaction equations; then discard this equation itself since it can no longer be relevant to the problem. Note, however, that this element should be used in Step 4 to formulate material-balance equations. In carrying out Step 2, a second elemental species not actually present may also be eliminated from the set of reaction equations. Only one of these elements simultaneously eliminated can form the basis for a material balance.

3. Repeat Step 2 until the set of reaction equations contains no elemental species presumed to be absent from the equilibrium mixture. Note that an element like carbon may be one of the actual reactants but absent at equilibrium.

4. Formulate a set of material balances based on the elements. Use the elements present in the equilibrium mixture plus those noted for this purpose in Step 2.

Where the six species present at equilibrium are NH_3, H_2O, CO_2, NH_4OH, $(NH_4)_2CO_3$, and NH_4HCO_3, reactions for formation from the elements are

$$N_2 + 3H_2 \rightarrow 2NH_3,$$

$$2H_2 + O_2 \rightarrow 2H_2O,$$

$$C + O_2 \rightarrow CO_2,$$

$$N_2 + 5H_2 + O_2 \rightarrow 2NH_4OH,$$

$$2N_2 + 8H_2 + 2C + 3O_2 \rightarrow 2(NH_4)_2CO_3,$$

and

$$N_2 + 5H_2 + 2C + 3O_2 \rightarrow 2NH_4HCO_3.$$

Since nitrogen does not appear at equilibrium, use the first equation to eliminate N_2. The set becomes

$$2H_2 + O_2 \rightarrow 2H_2O,$$

$$C + O_2 \rightarrow CO_2,$$

$$2H_2 + O_2 + 2NH_3 \rightarrow 2NH_4OH,$$

$$2H_2 + 2C + 3O_2 + 4NH_3 \rightarrow 2(NH_4)_2CO_3,$$

and

$$2H_2 + 2C + 3O_2 + 2NH_3 \rightarrow 2NH_4HCO_3.$$

Since hydrogen does not appear at equilibrium, use the first of the above set to eliminate H_2. The set becomes

$$C + O_2 \rightarrow CO_2,$$

$$H_2O + NH_3 \rightarrow NH_4OH,$$

$$C + O_2 + 2NH_3 + H_2O \rightarrow (NH_4)_2CO_3,$$

and

$$C + O_2 + NH_3 + H_2O \rightarrow NH_4HCO_3.$$

Since carbon does not appear at equilibrium, use the first of the above set to eliminate C. The set becomes

$$H_2O + NH_3 \rightarrow NH_4OH,$$

$$CO_2 + 2NH_3 + H_2O \rightarrow (NH_4)_2CO_3,$$

and

$$CO_2 + NH_3 + H_2O \rightarrow NH_4HCO_3.$$

Notice that O_2 was eliminated at the same time as C.

Three independent material-balance relations can also be formulated. These can be elemental balances on total N, total H, and total C. No independent balance on total O can be formulated.

Starting with the elemental composition of all species present at equilibrium, the digital computer can go through the above procedure to formulate the correct material balances and equilibrium relations. With a databank containing extended forms of Table 19.1 and Table 5.1, the computer can next solve for the equilibrium composition for a given initial composition and assumed values of fugacity coefficients. The problem requires specification of two other variables, such as the temperature and pressure. Alternatively, for an isolated system, the volume and internal energy might be specified, and for an adiabatic flow system, the pressure and enthalpy might be specified.

19.5 Chemical Equilibria

The thermodynamics of chemical reactions has already been treated in the preceding sections. High temperatures are frequently used to facilitate the kinetics (rates) of chemical reactions, and this makes the gas-phase reactions very important. Even so, catalysis is important. Catalysts can increase the rate without changing the equilibrium state and thereby permit an optimization between catalyst cost and treatment and the cost of operating at high temperatures. This brings us to heterogeneous chemical reactions, because heterogeneous catalysts have many practical advantages over homogeneous catalysts, including subsequent separation.

For heterogeneous reaction equilibria there is a short cut to obtain the equilibrium relationship. Write the reaction as

$$0 \to \sum_i s_i M_i, \tag{19.26}$$

where s_i is the stoichiometric coefficient for species i (positive for products and negative for reactants) and M_i is a symbol for the chemical formula of species i. The species can be from either adjacent phase or from the interface itself (as in detailed mechanisms of a reaction). Then the equilibrium relationship follows directly by replacing M_i by the chemical potential of the species.

$$0 \to \sum_i s_i \mu_i. \tag{19.27}$$

The chemical potentials can then be expressed with secondary-reference-state quantities as is appropriate for gas-phase species or liquid-phase species or dilute solutions (Chapters 19, 20, or 21).

Example 19.1 The water–gas shift reaction

$$CO(gas) + H_2O(gas) \rightleftharpoons CO_2(gas) + H_2(gas)$$

at 1000°F has a standard Gibbs-function change

$$\mu^*_{CO_2} + \mu^*_{H_2} - \mu^*_{CO} - \mu^*_{H_2O} = \Delta \tilde{G}^*$$

of −3850 Btu/lb-mol. A reaction mixture at this temperature and a total pressure of 1 atm consists initially of 1 mol of CO and 1 mol of H_2O. Calculate the Gibbs function of the system, relative to this initial condition, at 20, 40, ..., 100% conversion to CO_2 and H_2, and plot the values. Show that the lowest point on the curve agrees with that calculated from the equation

$$RT \ln K_p = -\Delta \tilde{G}^*,$$

where in this case

$$K_p = \frac{y_{CO_2} y_{H_2}}{y_{CO} y_{H_2O}}$$

if fugacity-coefficient corrections are ignored.

Solution: Let the fraction of conversion be denoted by C. Then, for one initial mole of CO and one of H_2O, the numbers of moles are

$$n_{CO} = 1 - C, n_{H_2O} = 1 - C, n_{CO_2} = C, n_{H_2} = C.$$

Furthermore, the total number of moles present is always 2, so the mole fractions are these above values divided by 2.

Since $G = \sum_i n_i \mu_i$ (Eq. 14.12) and the chemical potentials can be expressed as

$$\mu_i = \mu_i^*(T) + RT \ln py_i \phi_i$$

(Eq. 18.5), the Gibbs function relative to the initial condition is

$$\Delta G = \sum_i n_i \mu_i - \sum_i n_i \mu_i \text{ (initial)}$$

$$= (1-C)\left(\mu_{CO}^* + RT \ln p\frac{1-C}{2}\right) + (1-C)\left(\mu_{H_2O}^* + RT \ln p\frac{1-C}{2}\right)$$

$$+ C\left(\mu_{CO_2}^* + RT \ln pC/2\right) + C\left(\mu_{H_2}^* + RT \ln pC/2\right)$$

$$- \left(\mu_{CO}^* + RT \ln p/2\right) - \left(\mu_{H_2O}^* + RT \ln p/2\right)$$

$$= C\Delta\tilde{G}^* + 2RT[(1-C)\ln(1-C) + C \ln C],$$

where fugacity-coefficient corrections have been ignored. The last term is zero when $C = 0$ and again when $C = 1$. The first term is equal to zero when $C = 0$ and equal to $\Delta\tilde{G}^*$ when $C = 1$. Thus, $\Delta\tilde{G}^*$ is the value of ΔG when $C = 1$, at complete conversion, not at equilibrium.

To make the graph, we compute the dimensionless quantity

$$\frac{\Delta G}{RT} = C\frac{\Delta\tilde{G}^*}{RT} + 2(1-C)\ln(1-C) + 2C \ln C,$$

where

$$\frac{\Delta\tilde{G}^*}{RT} = \frac{-3850}{1.987 \times 1459.6} = -1.3274.$$

The results are as follows:

C	ΔG/RT	C	ΔG/RT	C	ΔG/RT
0	0	0.5	-2.0500	0.7	-2.1509
0.1	-0.7829	0.6	-2.1425	0.8	-2.0627
0.2	-1.2663	0.65	-2.1577	0.9	-1.8448
0.3	-1.6199	0.66	-2.1582	1.0	-1.3274
0.4	-1.8770	0.67	-2.1577		

Figure 19.3 shows the shape of the curve.

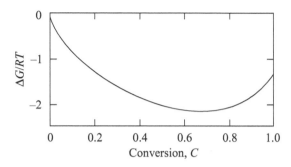

Figure 19.3 Gibbs function at 1000°F for a stoichiometric mixture of CO and H_2O, for varying degree of conversion according to the water–gas shift reaction.

On the other hand, with the equilibrium-constant expression, we have

$$\frac{p_{H_2}p_{CO_2}}{p_{H_2O}p_{CO}} = \frac{y_{H_2}y_{CO_2}}{y_{H_2O}y_{CO}} = K_p = e^{-\Delta\tilde{G}^*/RT} = 3.77 \, .$$

The total pressure conveniently cancels. Substitution for the mole fractions gives

$$\frac{C^2}{(1-C)^2} = K_p$$

or

$$C = \frac{\sqrt{K_p}}{1+\sqrt{K_p}} = \frac{\sqrt{3.77}}{1+\sqrt{3.77}} = 0.66 \, ,$$

in good agreement with the minimum in the graph of $\Delta G/RT$ versus C.

Example 19.2 Calculate the equilibrium conversion to ammonia of a stoichiometric mixture of nitrogen and hydrogen as a function of temperature and pressure in the ranges 25°C to 600°C and 1 atm to 1000 atm. You may neglect fugacity-coefficient corrections.

Solution: First we must calculate the equilibrium constant K as a function of temperature. Express the change of heat capacity on reaction on the basis of Eq. 5.6 as

$$\Delta C_p^* = \Delta a + T\Delta b + \Delta c /T^2 \, ,$$

where, from Table 5.1,

$$\Delta a = \sum_i v_i a_i = -12.17 \text{ cal/mol-K},$$

$$\Delta b = 0.00876 \text{ cal/K}^2\text{-mol},$$

and $$\Delta c = -9.8 \times 10^4 \text{ cal-K/mol}.$$

Integration with respect to T gives the enthalpy change of the reaction as it depends on temperature:

$$\Delta H^* = \Delta H^*_{298} + \Delta a(T - T_o) + \frac{\Delta b}{2}(T^2 - T_o^2) - \Delta c\left(\frac{1}{T} - \frac{1}{T_o}\right),$$

where ΔH^*_{298} is the value (−22.04 kcal/mol) at 25°C obtained from Table 19.1 and T_o represents this reference temperature (298.15 K).

ΔH^* can now be divided by RT^2 and integrated again to obtain ΔG^* and the equilibrium constant K according to Eq. 19.22. The results are plotted in Fig. 19.2. Multiplication by p^2 gives K_y according to Eq. 19.14, and the equilibrium conversion can be determined from Eq. 19.18 or Fig. 19.1.

Figure 19.4 gives the equilibrium conversion as a function of temperature and pressure. Since the reaction is exothermic, a high degree of conversion is favored by low temperatures. Since the number of moles of product is less than the number of moles of reactants, high pressures favor the conversion. Notice that the rather extensive information summarized in Fig. 19.4 is based on just a few entries in Table 19.1 and Table 5.1. Values of log K can also be found tabulated against temperature in Ref. [3].

To understand the conditions of industrial ammonia synthesis, one must also realize that the reaction is very slow at low temperatures. Thus, high temperatures are used to increase the rate of the reaction, and a heterogeneous catalyst such as Fe reduced in place from Fe_3O_4 and containing K_2O is used as well. The operating pressure is raised to compensate for the smaller equilibrium constant at higher temperatures. The ammonia synthesis may thus be carried out on a catalyst at 350 atm and 500°C. The reaction is not carried to completion, nor even to the equilibrium conversion at the reactor conditions. Ammonia is removed from the product stream, and unreacted nitrogen and hydrogen are recycled. Clearly, engineers have to optimize the reactor conditions while taking into account equilibrium conditions, catalysis behavior, and separation costs.

Inclusion of the fugacity coefficients will modify Fig. 19.4 only in detail. Problem 18.10 suggests that ϕ_{NH_3} may be about 0.37 at 150°C and 350 atm. This will displace the curve upward slightly. However, the fugacity coefficient will increase toward unity at higher temperatures. The critical temperature and pressure of ammonia are 132.5°C and 112.5 atm. Thus, liquid ammonia may separate out at low temperatures and high pressures, and this can be used in the process of recycling the unreacted hydrogen and nitrogen.

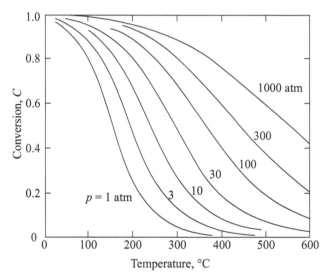

Figure 19.4 Temperature and pressure dependence of the equilibrium conversion to ammonia from a stoichiometric mixture of nitrogen and hydrogen.

Example 19.3 Develop a computer program for determining the equilibrium composition when several reactions must be equilibrated simultaneously. Consider specifically the reforming of methane according to the reactions

$$CH_4 + H_2O \rightleftharpoons CO + 3H_2, K_{p1} = 0.41 \text{ atm}^2,$$

$$CH_4 + 2H_2O \rightleftharpoons CO_2 + 4H_2, K_{p2} = 1.09 \text{ atm}^2,$$

where the equilibrium constants are given at 1100°F.

Solution: If we neglect fugacity coefficients, that is, $\phi_i = 1$, then the equilibrium expressions are given by:

$$\frac{K_{p1}}{p^2} = \frac{y_{CO}\,y_{H_2}^3}{y_{CH_4}\,y_{H_2O}} \quad \text{and} \quad \frac{K_{p2}}{p^2} = \frac{y_{CO_2}\,y_{H_2}^4}{y_{CH_4}\,y_{H_2O}^2}.$$

Instead of using the *extent* of each reaction, we can express the stoichiometry in the form of material balances as follows:

(a) Carbon balance

$$n_{CO} + n_{CO_2} + n_{CH_4} = n_{CO}^o + n_{CO_2}^o + n_{CH_4}^o,$$

(b) Hydrogen balance

$$2n_{CH_4} + n_{H_2} + n_{H_2O} = 2n_{CH_4}^o + n_{H_2}^o + n_{H_2O}^o,$$

(c) Oxygen balance

$$n_{CO} + 2n_{CO_2} + n_{H_2O} = n_{CO}^o + 2n_{CO_2}^o + n_{H_2O}^o,$$

where n_j and n_j^o are the numbers of moles of species j present at equilibrium and initially, respectively. The total number of moles n at equilibrium is the sum for all the species present:

$$n = n_{CO} + n_{CO_2} + n_{CH_4} + n_{H_2} + n_{H_2O}.$$

We now decide to regard the five mole numbers in this equation as the unknowns to be solved for, rather than the mole fractions. Consequently, we rewrite the equilibrium relations in the form

$$\frac{K_{p1}}{p^2} = \frac{n_{CO}\,n_{H_2}^3}{n_{CH_4}\,n_{H_2O}\,n^2} \quad \text{and} \quad \frac{K_{p2}}{p^2} = \frac{n_{CO_2}\,n_{H_2}^4}{n_{CH_4}\,n_{H_2O}^2\,n^2}.$$

We intend to solve the five governing equations simultaneously, using a general matrix-inversion program (see program listing). However, the equilibrium relationships are nonlinear and must be linearized by the following procedure. We assume that we have a set constituting a trial solution, which is fairly close to the correct answer and which will be denoted by an asterisk (*). Any function f that depends on the unknowns can be expanded in a Taylor expansion about the trial values:

$$f(n_1, n_2, \ldots) = f(n_1^*, n_2^*, \ldots) + \sum_j \left. \frac{\partial f}{\partial n_j} \right|_{n_1^*, n_2^*, \ldots} \Delta n_j + \cdots,$$

where $\Delta n_j = n_j - n_j^*$. Terms that are not constant or linear in the unknowns are dropped in this and subsequent manipulations of the equations. The equations are thus all linearized, and the matrix-inversion procedure can be applied. The solution obtained should be much closer to the correct answer and can now be used as the trial solution. The procedure is then repeated (a process called *iteration*) until no further change in the answer is observed (or *convergence* is achieved). The solution should now satisfy the original, nonlinear equations.

The preceding solution method is a generalization of the so-called Newton–Raphson method frequently applied to a single nonlinear equation with a single unknown, and it is a very powerful way to attack a large set of coupled, nonlinear equations. It works beautifully when the trial solution is close to the correct answer, but precautions may be necessary to get it started in the right direction. We shall build some of these into our procedure.

First, to put all of our equilibrium expressions into a standard form that is easy to handle, take the logarithm of all such equations. We have here

$$\ln\left(\frac{K_{p1}}{p^2}\right) = \ln n_{CO} + 3\ln n_{H_2} - \ln n_{CH_4} - \ln n_{H_2O} - 2\ln n$$

and

$$\ln\left(\frac{K_{p2}}{p^2}\right) = \ln n_{CO_2} + 4\ln n_{H_2} - \ln n_{CH_4} - 2\ln n_{H_2O} - 2\ln n .$$

Expansion of a function like $\ln n_{CO}$ in a Taylor series yields

$$\ln n_{CO} = \ln n_{CO}^* + \frac{\Delta n_{CO}}{n_{CO}^*} + \cdots,$$

and the expansion of a term like $\ln n$ yields

$$\ln n = \ln n^* + \sum_j \frac{\Delta n_j}{n^*} + \cdots .$$

The linearized form of the first equilibrium expression now becomes

$$\ln\left[\frac{K_{p1}n^*_{CH_4}n^*_{H_2O}(n^*)^2}{p^2n^*_{CO}(n^*_{H_2})^3}\right] = \frac{\Delta n_{CO}}{n^*_{CO}} + \frac{3\Delta n_{H_2}}{n^*_{H_2}} - \frac{\Delta n_{CH_4}}{n^*_{CH_4}}$$

$$-\frac{\Delta n_{H_2O}}{n^*_{H_2O}} - 2\sum_j \frac{\Delta n_j}{n^*}. \tag{A}$$

The matrix form into which the linearized equations are to be cast is

$$\sum_j B_{i,j}\Delta n_j = G_i.$$

To see just how the computer program is developed, one should follow the statements just below statement 3, where the first equilibrium relationship is programmed. G_i represents the constant part of the equation and should take the form (for $i = 1$)

$$G_i = \ln\left[\frac{K_{p1}n^*_{CH_4}n^*_{H_2O}(n^*)^2}{p^2n^*_{CO}(n^*_{H_2})^3}\right]. \tag{B}$$

$B_{i,j}$ represents the coefficient of Δn_j in the linearized form of equation i. The last term in Eq. A generates a contribution $-2/n^*$ to the $B_{i,j}$ value for each j in equation i. The remaining terms give coefficients explicitly for each chemical species. Thus, we have

$$B_{i,CO} = \frac{1}{n^*_{CO}} - \frac{2}{n^*}, \qquad B_{i,H_2} = \frac{3}{n^*_{H_2}} - \frac{2}{n^*},$$

$$B_{i,CH_4} = \frac{-1}{n^*_{CH_4}} - \frac{2}{n^*}, \quad \text{and} \quad B_{i,H_2O} = \frac{-1}{n^*_{H_2O}} - \frac{2}{n^*}.$$

Although CO_2 does not appear explicitly, the last term in Eq. A still leaves us with

$$B_{i,CO_2} = \frac{-2}{n^*}.$$

The notation in the computer program is designed to make it obvious what is being put into the equations. NT in the listing represents n or n^*.

Although the above procedure may seem cumbersome to master, its widespread utility makes the effort well worthwhile. The readers should follow through with the second equilibrium relationship and verify that the equation for $i = 2$ is programmed correctly in the computer listing. The equations for the carbon balance, the hydrogen balance, and the oxygen balance were linear to begin with, but they need to be expressed in terms of Δn_j instead of n_j. The readers should be able to see how the material balances are implemented in the computer program. The initial mole numbers are denoted NI in the listing.

The equations are now linear, and after the B and G arrays have been set up, the solution is obtained by calling the matrix-inversion subroutine. The solution Δn_j comes back in the G array. Before these are added to the n_j^* array, a limit is placed on them to prevent the new value of n_j from decreasing by more than a factor of 10^3. Thus, no negative numbers can be generated, and the calculation method is quite stable. We can see that starting guesses of zero or negative numbers should be avoided. In fact, if the starting guesses are on the high side, the program should never fail. Maximum possible values for each species can always be obtained from the material balances. (An extremely low starting guess for a minor species will result in slow convergence because the mole fraction for this species will increase by only a modest amount for each iteration. See Problem 19.6.)

In the program, convergence is checked after statement 4. The check should be made on a sensitive variable, or on all the variables. If convergence has not been achieved, the program loops back to statement 2 and begins a new iteration. The printed results for specified pressure, equilibrium constants, and initial mole numbers show convergence after six iterations.

The readers should be able to modify this fundamental computer program to treat different chemical equilibria with different stoichiometrics. Incidentally, we have solved for the change Δn_j at each iteration because this gives better accuracy for some difficult problems than a direct matrix solution for the unknown n_j itself. It also makes it easier to check the programming since G_i is now a statement of the original equation. See Eq. B.

Methane reforming is an important process for producing hydrogen for ammonia synthesis and other uses. Hydrogen is produced this way for fuel cells with phosphoric acid electrolyte because direct electrochemical reaction of a hydrocarbon feed is more difficult. In practice, reforming is carried out at a temperature of about 800°C and at pressures from 1 to 30 atm, depending on the application, and 3 to 5 mol of water are used per mole of methane.

Subsequently, more steam is added, the temperature is dropped, and CO is shifted to H_2 on a catalyst specific for the water–gas shift reaction. The conditions for this example were chosen more to have equilibrium constants near unity.

```
C       PROGRAM GAS(INPUT,OUTPUT)
        IMPLICIT REAL*8(A-H,0-Z)
        REAL*8 N,NI,NT,NIT
        DIMENSION B(8,8),N(8),G(8),Y(8),NI(8)
        COMMON B,G
 100 FORMAT (8X,I2,1X,6E11.4)
 101 FORMAT (5F10.0)
 199 FORMAT (26H THIS RUN DID NOT CONVERGE)
 200 FORMAT(1H1,8X,3HP=,F5.2,2X,5HKP1=,E12.5,2X,5HKP2=,E12.5)
 201 FORMAT(1H0,14X,3HTOT,9X,2HCO,8X,3HCO2,8X,3HCH4,9X,2HH2,8X,3HH2O/
    1 33X,16HNUMBERS OF MOLES)
 202 FORMAT (32X, 14HMOLE FRACTIONS/(22X,5E11.4))
 203 FORMAT (3X, 8H INITIAL, 6E11.4)
C       SUBSCRIPTS FOR THE UNKNOWNS
        JCO = 1
        JCO2 = 2
        JCH4 = 3
        JH2 = 4
        JH2O = 5
        JMAX=5
C       PRESSURE AND EQUILIBRIUM CONSTANTS
        READ 101, P, QKP1, QKP2
        PRINT 200, P, QKP1, QKP2
C       INITIAL NUMBERS OF MOLES
        READ 101, (NI(J), J=1, JMAX)
        NIT=0.0D0
        DO 1 J = 1, JMAX
      1 NIT=NIT + NI (J)
        PRINT 201
        PRINT 203, NIT, (NI(J), J = 1, JMAX)
C       INITIAL GUESSES
        N(JCO)=0.19D0
        N(JCO2)=0.59D0
        N(JCH4)=4.2D0
        N(JH2)=2.94D0
        N(JH2O)=0.8D0
        JCOUNT=0
C       ZERO THE MATRIX AND CALCULATE TOTAL
        NUMBER OF MOLES
      2 JCOUNT=JCOUNT + 1
        NT=0.0D0
```

```
          DO 3 I=1, JMAX
          G(I)=0.0D0
          NT=NT+N(I)
          DO 3 J=1, JMAX
        3 B(I,J)=0.0D0
          A=2.0D0/NT
C         EQUILIBRIUM FOR THE REACTION, CH4 + H2O = CO + 3H2
          I=1
          G(I)=DLOG(QKP1*N(JCH4)*N(JH2O)*(NT/P)**2/N(JCO)/N (JH2)**3)
          B(I,JCO)=1.0D0/N(JCO)-A
          B(I,JCO2) = -A
          B(I,JCH4) = -1.0D0/N(JCH4)-A
          B(I,JH2) = 3.0D0/N(JH2)-A
          B(I,JH2O) = -1.0D0/N(JH2O)-A
C         EQUILIBRIUM FOR THE REACTION, CH4 + 2H2O = CO2 + 4H2
          I=2
          G(I)=DLOG (QKP2*N(JCH4)*(N(JH2O)*NT/P)**2/N(JCO2)/N(JH2)**4)
          B(I,JCO)=-A
          B(I,JCO2)=1.0D0/N(JCO2)-A
          B(I,JCH4) = -1.0D0/N(JCH4)-A
          B(I,JH2)=4.0D0/N(JH2)-A
          B(I,JH2O)= -2.0D0/N(JH2O)-A
C         CARBON BALANCE
          I=3
          B(I,JCO)=1.0D0
          B(I,JCO2)=1.0D0
          B(I,JCH4)=1.0D0
C         HYDROGEN BALANCE
          I=4
          B(I,JH2)=1.0D0
          B(I,JH2O)=1.0D0
          B(I,JCH4)=2.0D0
C         OXYGEN BALANCE
          I=5
          B(I,JCO)=1.0D0
          B(I,JCO2)=2.0D0
          B(I,JH2O)=1.0D0
          DO 4 I=3, JMAX
          DO 4 J=1, JMAX
        4 G(I)=G(I)+B(I, J)*(NI(J)-N(J))
          CALL MATINV(JMAX, 1, DETERM)
          KERR=0
          ERR=1.0D-6
```

```
C     CHECK CONVERGENCE AND MAKE CORRECTIONS TO N(J)
      DO 5 J=1, JMAX
      IF (DABS(G(J)). GT.ERR*N(J)) KERR=KERR+1
      IF (G(J).LT.-0.999D0*N(J)) G(J)=-0.999D0*N(J)
      N(J)=N(J)+G(J)
    5 Y(J)=N(J)/NT
      PRINT 100, JCOUNT,NT,(N(J), J=1, JMAX)
      IF(JCOUNT.LT.20 .AND. KERR.NE.0) GO TO 2
      IF(KERR.NE.0) PRINT 199
      PRINT 202, (Y(J), J=1,JMAX)
      STOP
      END
      SUBROUTINE MATINV (N,M,DETERM)
      IMPLICIT REAL*8(A-H, O-Z)
      COMMON B, D
C     MATRIX INVERSION WITH ACCOMPANYING SOLUTION OF LINEAR EQUATIONS.
      DIMENSION B(8,8),D(8,1),ID(8)
      DETERM=1.0D0
      DO 1       I=1,N
    1 ID(I)=0
      DO 18 NN=1,N
      BMAX=1.1D0
      DO 6 I=1,N
      IF(ID(I).NE.0) GO TO 6
      BNEXT=0.0D0
      BTRY=0.0D0
      DO 5 J=1,N
      IF(ID(J).NE.0) GO TO 5
      IF(DABS(B(I,J)).LE.BNEXT) GO TO 5
      BNEXT=DABS(B(I,J))
      IF(BNEXT.LE.BTRY) GO TO 5
      BNEXT=BTRY
      BTRY=DABS(B(I,J))
      JC=J
    5 CONTINUE
      IF(BNEXT.GE.BMAX*BTRY) GO TO 6
      BMAX=BNEXT/BTRY
      IROW=I
      JCOL=JC
    6 CONTINUE
      IF(ID(JC).EQ.0) GO TO 8
      DETERM=0.0D0
      RETURN
```

```
   8 ID(JCOL)=1
     IF(JCOL.EQ.IROW) GO TO 12
     DO 10 J=1,N
     SAVE=B(IROW,J)
     B(IROW,J)=B(JCOL,J)
  10 B(JCOL,J)=SAVE
     DO 11 K=1,M
     SAVE=D(IROW,K)
     D(IROW, K)=D(JCOL,K)
  11 D(JCOL,K)=SAVE
  12 F=1.0D0/B(JCOL,JCOL)
     DO 13 J=1,N
  13 B(JCOL,J)=B(JCOL,J)*F
     DO 14 K=1,M
  14 D(JCOL,K)=D(JCOL,K)*F
     DO 18 I=1,N
     IF(I .EQ.JCOL) GO TO 18
     F=B(I,JCOL)
     DO 16 J=1,N
  16 B(I,J)=B(I, J)-F*B(JCOL,J)
     DO 17 K=1,M
  17 D(I,K)=D(I,K)-F*D(JCOL,K)
  18 CONTINUE
     RETURN
     END
$ENTRY
2.000         0.410       1.090
0.000         0.000       5.000       0.000       7.000

P = 2.00     KP1= 0.41000D 00      KP2= 0.10900D 01
```

	TOT	CO	CO_2	CH_4	H_2	H_2O
			NUMBERS OF MOLES			
INITIAL	0.1200D 02	0.0000D 00	0.0000D 00	0.5000D 01	0.0000D 00	0.7000D 01
1	0.8720D 01	4720D 00	0.1567D 01	0.2961D 01	0.7684D 01	0.3394D 01
2	01608D 02	0.7532D 00	0.1203D 01	0.3044D 01	0.7072D 01	0.3841D 01
3	0.1591D 02	0.8209D 00	0.1173D 01	3006D 01	0.7155D 01	0.3833D 01
4	0.1599D 02	0.8232D 00	0.1172D 01	0.3005D 01	0.7157D 01	0.3833D 01
5	0.1599D 02	0.8232D 00	0.1172D 01	0.3005D 01	0.7157D 01	0.3833D 01
6	0.1599D 02	0.8232D 00	0.1172D 01	0.3005D 01	0.7157D 01	0.3833D 01
			MOLE FRACTIONS			
		0.5148D-01	0.7329D-01	0.1879D 00	0.4476D 00	0.2397D 00

Example 19.4 By an orderly successive-approximation procedure, without a computer, determine the equilibrium composition for

the combustion of propane with a 20% excess of air. It is desired to estimate the amount of any H_2, CO, NO, and NO_2 that might be formed if the flue gas can be assumed to be equilibrated at 1000°F and 1 atm.

Solution: The complete combustion of propane requires 5 mol of oxygen:

$$C_3H_8 + 5O_2 \rightarrow 3CO_2 + 4H_2O.$$

A 20% excess of air means that 6 mol of oxygen are actually used, and these will bring an additional $6 \times (0.79/0.21) = 22.57$ mol of nitrogen into the combustion system. With a basis of 1 mol of propane, the total moles of each element in the system will be 3 of carbon, 8 of hydrogen, 12 of oxygen, and 45.14 of nitrogen. The major species in the equilibrium mixture will be O_2, H_2O, CO_2, and N_2, and these are determined principally by the material balances. The minor species will be H_2, CO, and NO_2, and these will be determined principally by the equilibrium expressions, for which we can obtain the values (at 1000°F)

$$N_2 + O_2 \rightleftarrows 2\,NO, \quad K = 4.57 \times 10^{-11},$$

$$2H_2 + O_2 \rightleftarrows 2\,H_2O, \quad K = 1.389 \times 10^{26} \text{ atm}^{-1},$$

$$2O_2 + N_2 \rightleftarrows 2\,NO_2, \quad K = 1.563 \times 10^{-11} \text{ atm}^{-1},$$

$$CO + H_2O \rightleftarrows CO_2 + H_2, \quad K = 3.749.$$

We can begin by neglecting the minor species and determining the major species from material balances. The nitrogen balance gives

$$n_{N_2} = 45.14 / 2 = 22.57,$$

and the hydrogen balance gives

$$n_{H_2O} = 8/2 = 4.$$

For carbon, we have

$$n_{CO_2} = 3.$$

Finally, the oxygen balance yields

$$n_{O_2} = \left(12 - n_{H_2O} - 2n_{CO_2}\right)/2 = (12 - 4 - 2 \times 3)/2 = 1.0,$$

corresponding to the extra mole of O_2 added to provide the 20% excess of air. The total number of moles can be estimated to be

$$n = 22.57 + 4 + 3 + 1 = 30.57.$$

Next, the minor species can be estimated from the equilibrium relations. The moles of nitric oxide amount to

$$n_{NO} = \left(4.57 \times 10^{-11} n_{N_2} n_{O_2}\right)^{1/2} = 3.21 \times 10^{-5}.$$

For hydrogen, we obtain

$$n_{H_2} = \frac{n_{H_2O}}{(1.389 \times 10^{26} p\, n_{O_2}/n)^{1/2}} = 1.88 \times 10^{-12}.$$

Nitrogen dioxide amounts to

$$n_{NO_2} = n_{O_2} (1.563 \times 10^{-11} p\, n_{N_2}/n)^{1/2} = 3.40 \times 10^{-6},$$

and the number of moles of carbon monoxide is

$$n_{CO} = \frac{n_{CO_2} n_{H_2}}{3.749\, n_{H_2O}} = 3.75 \times 10^{-13}.$$

Second approximations could be obtained by going again through the preceding sequence of calculations, with the best available values for all the mole numbers except that one which is being calculated at the moment. For example, the nitrogen balance would now yield

$$n_{N_2} = (45.14 - n_{NO} - n_{NO_2})/2 = 22.56998.$$

In this example, the minor species have such low equilibrium values that a second iteration is hardly necessary. The equilibrium mole fractions now are

$$y_{N_2} = 0.7383, \qquad y_{NO} = 1.05 \times 10^{-6},$$

$$y_{H_2O} = 0.1308, \qquad y_{H_2} = 6.15 \times 10^{-14},$$

$$y_{CO_2} = 0.0981, \qquad y_{CO} = 1.23 \times 10^{-14},$$

$$y_{O_2} = 0.0327, \quad \text{and}\ y_{NO_2} = 1.11 \times 10^{-7}.$$

In some problems of equilibrium composition, a little thought will reveal a successive-approximation scheme, which will converge in a few enough iterations so as to be suitable for hand calculation. For repetitive calculations with variations of the input parameters (like the excess air, the pressure, or the temperature), one would probably want to develop a computer program based on this scheme. For those who choose to devote their mental effort elsewhere, the general approach of Example 19.3 can be recommended. See also Problems 19.5 and 19.7.

Problems

19.1 What is the equilibrium concentration of NO_2 in the atmosphere at 298 K? For the reaction

$$\frac{1}{2}N_2 + O_2 \rightarrow NO_2,$$

$$\Delta H^*_{298} = 7930 \text{ cal/mol}$$

and

$$\Delta G^*_{298} - 12,260 \text{ cal/mol}.$$

Compare your result with the value considered to constitute smog conditions: 0.25 parts per million.

19.2 Prepare a semi-logarithmic graph of K versus $1/T$ for the reactions given below. Use only the data in Table 19.1, and assume that ΔH^* is independent of temperature. The information in Problem 19.3 will give some insight into the accuracy of this approximation.

   a. The water–gas shift reaction.

   b. Ammonia synthesis.

   c. Hydration of ethylene to form ethanol.

   d. Synthesis of methane from CO_2 and H_2.

19.3 Estimate K at 800 K for the reactions given below. Use only the data in Table 19.1, and assume that ΔC^*_p is independent of temperature. For comparison with your results, actual K values at 800 K are given. Since these actual values account for the temperature dependence of ΔC^*_p, they may differ somewhat from your results.

   a. The water–gas shift reaction.

   $$CO + H_2O \rightleftharpoons CO_2 + H_2, \quad K = 4.24.$$

   b. Ammonia synthesis.

   $$N_2 + 3H_2 \rightleftharpoons 2NH_3,$$

   $$K = 8.99 \times 10^{-6} \text{ atm}^{-2}.$$

c. Hydration of ethylene to form ethanol.
$$C_2H_4 + H_2O \rightleftharpoons C_2H_5OH \, ,$$
$$K = 0.210 \text{ atm}^{-1}.$$

d. Synthesis of methane from CO_2 and H_2.
$$CO_2 + 4H_2 \rightleftharpoons CH_4 + 2H_2O \, ,$$
$$K = 7.83 \text{ atm}^{-2}.$$

19.4 One pound of coal having the ultimate analysis (by weight)

Moisture	2.9%
Carbon	73.8
Available H	3.8
Sulfur	1.1
Nitrogen	1.4
Corrected ash	9.8
Combined H_2O	7.2

is burned with 8 lb of dry air. It is desired to treat the equilibria in the product flue gas, which is considered to contain possibly carbon monoxide, carbon dioxide, sulfur dioxide, sulfur trioxide, oxygen, nitrogen, and water.

a. Write down a complete set of independent chemical reaction equations, which will form a basis for determining the equilibrium composition of the system.

b. Indicate the material-balance relations that could be written in order to complete the solution for the equilibrium composition.

c. If the temperature and pressure of the flue gas are 1000°F and 1 atm, indicate that you have enough equations to solve for the equilibrium composition if you had the time and adequate computation facilities.

19.5 Propane is burned with a 20% excess of air. After combustion, hydrogen and carbon monoxide are added in sufficient quantity to consume the excess oxygen and also remove most of the nitrogen oxides that have been formed. The temperature is 1000°F, and the pressure is 1 atm. Calculate the equilibrium composition as a function of the amounts of added hydrogen and carbon monoxide. For each mole of propane burned, take the added moles of hydrogen to be $1 + \varepsilon$ and the added moles of carbon monoxide to be $1 + \delta$. Assume that the species present are NO_2, NO, O_2, N_2, H_2, CO, CO_2, and H_2O. Prepare a computer solution using equal values of ε and δ between 1 and 2 of your own choosing. If you have the time, try some other (equal) values so that you obtain an idea of the effect of the added hydrogen and carbon monoxide.

a. Due 2 days after the assignment of the problem, submit a statement of the independent chemical reactions possible in the mixture. Also express the required number of material balances on the basis of 1 mol of propane.

   Use mole numbers (not mole fractions) as the composition variables in this problem. (Example 19.4 may provide some useful hints, but you should try to work out your answer independently of this source.)

b. Obtain a matrix-inversion program from the internet or use computation programs you are familiar with. Estimate equilibrium constants at 1000°F for the chemical reactions you concluded are important.

c. Working independently, make relatively minor modifications to the computer program in Example 19.3. Try to preserve the input-output and iterative looping parts of the program so that you can concentrate on formulating the material-balance relations and equilibrium relations, the latter being properly linearized first. With this program, obtain numerical values for the resulting equilibrium composition.

19.6 Treat the solution to the equation $\ln y = 0$ by the Newton–Raphson method, ignoring for the moment the fact that the solution is $y = 1$.

a. Suppose that we make a starting guess of $y^* = e^{-50}$. Carry through two or three iterations of the calculation procedure.

b. Estimate how many iterations will be required to achieve convergence.

c. Suppose we make a starting guess of $y^* = e^2$. What will happen on the first iteration?

19.7 After one has obtained a complete solution to Problem 19.5, how difficult would it be to go back and modify the computer program to include the possibility of forming an additional species, ammonia, under the reducing conditions prevailing when excess hydrogen and carbon monoxide have been added to the flue gas?

19.8 One mole of carbon monoxide and two moles of hydrogen are initially charged into a vessel where the temperature is maintained at 700°F and the pressure is maintained at 250 bar. Calculate and plot the Gibbs function of the system as a function of the *extent of reaction* if it is presumed that CO and H_2 can react to form methanol according to the reaction

$$CO + 2H_2 \rightarrow CH_3OH.$$

At the operating temperature, the following data are given

$$\mu_{H_2}^* = -4.3 \, kJ/mol,$$

$$\mu_{CO}^* = -172.5 \, kJ/mol,$$

$$\mu_{CH_3OH}^* = -125.3 \, kJ/mol,$$

and ideal-gas behavior can be assumed. Pay particular attention to any minimum in the curve.

19.9 In one step in the manufacture of nitric acid, ammonia is burned with oxygen to produce nitric oxide and water, the combustion temperature being 920°C and the pressure being about 6 atm. If the reaction is carried out isothermally in a continuous flow reactor (with reactants preheated to the reactor temperature), how much heat will be absorbed or released per mole of ammonia reacted?

19.10 The second step in the manufacture of nitric acid involves the further oxidation of nitric oxide to nitrogen dioxide at a temperature of 900 K and a pressure of 6 atm. For the reaction

$$2NO + O_2 \rightleftharpoons 2NO_2,$$

the equilibrium constant K at this temperature is 0.07055 atm^{-1}. For a stoichiometric ratio of NO and O_2, determine the equilibrium conversion of nitric oxide and the equilibrium composition. Use the method of the extent of reaction, and assume that all fugacity coefficients are equal to unity.

19.11 Nitric acid reacts with nitric oxide in the gas phase to form nitrogen dioxide and water at a temperature of 800 K and a pressure of 4 atm. Determine the equilibrium constants K

and K_y for this reaction under these conditions. Assume that all fugacity coefficients are equal to unity. Be sure to indicate the dimensions of these quantities and that you have defined them by specifying the stoichiometry of the reaction you are describing. Does K or K_y (or both) depend on pressure or on temperature?

19.12 Hydrogen and carbon monoxide, initially in the ratio of 1 mol of hydrogen to 1 mol of CO, react to produce methanol. Determine the mole fraction of CO as a function of the extent of reaction.

19.13 A gas stream in a nitric acid plant is suspected to contain simultaneously NH_3, NO, NO_2, HNO_3, N_2, O_2, and H_2O. Give a complete set of independent reactions that might be used to describe chemical equilibrium in this system.

19.14 Calculate the equilibrium constant for the reaction

$$2H_2 + O_2 \rightleftharpoons 2H_2O$$

at a temperature of 1000°F. Should the equilibrium constant increase or decrease as the temperature is increased above 1000°F?

19.15 For small-scale research purposes, pure hydrogen is sometimes obtained by decomposition of ammonia. What will be the equilibrium mole fraction of hydrogen if ammonia is equilibrated over a suitable catalyst at 250°C and 2 atm?

19.16 One mole of ammonia and one mole of oxygen are introduced at 6 atm into a constant-volume chamber at 920°C. While the chamber is maintained at this temperature, the oxidation of ammonia to nitric oxide is allowed to proceed. Show how to treat the pressure variation in this system as the reaction proceeds; in particular, develop an expression for the pressure as a function of the extent of reaction if the mixture can be assumed to behave as an ideal gas. Substitute this result into the relationship for the chemical equilibrium so as to obtain a single equation that should be solved for the extent of reaction.

19.17 Are the following chemical reactions independent? Indicate your reasoning.

$$4H_2S + 2SO_2 \rightleftharpoons 3S_2 + 4H_2O$$
$$3H_2S + SO_2 \rightleftharpoons 2S_2 + 3H_2O$$

$$H_2O + SO_2 \rightleftharpoons H_2SO_3$$
$$H_2S + SO_3 \rightleftharpoons H_2S_2O_3$$

19.18 The second step in the manufacture of nitric acid involves the further oxidation of nitric oxide to nitrogen dioxide at a temperature of 900 K and a pressure of 6 atm. Determine the equilibrium constants K and K_y for this reaction under these conditions. Assume that all fugacity coefficients are equal to unity. Be sure to indicate the dimensions of these quantities and that you have defined them by specifying the stoichiometry of the reaction you are describing. Does K or K_y (or both) depend on pressure or on temperature?

19.19 Nitric acid can react with nitric oxide in the gas phase according to the reaction

$$2\,HNO_3 + NO_3 \rightleftharpoons 3NO_2 + H_2O\cdot$$

At 800 K, the equilibrium constant K for this reaction is 1.742 × 10^6 atm. For a stoichiometric ratio of HNO_3 and NO at a pressure of 4 atm, determine the equilibrium conversion of nitric acid and the equilibrium composition. Use the method of the extent of reaction.

19.20 Ethylene is an important intermediate in the chemical industry. Investigate its production by the thermal dehydrogenation of ethane by calculating the equilibrium constants K and K_y at the operating conditions of 1520°F and 1 atm and with a feed of 5 mol of water for each mole of ethane entering the reactor. Assume that all fugacity coefficients are unity. Be sure to indicate the dimensions of K and K_y and that you have defined them by specifying the stoichiometry of the reaction you are describing. Does K or K_y (or both) depend on pressure or on temperature?

19.21 For the thermal dehydrogenation of ethane at 1520°F according to the reaction

$$C_2H_6 \rightleftharpoons C_2H_4 + H_2\cdot$$

the equilibrium constant is $K = 2.45$ atm. Determine the equilibrium conversion and the equilibrium composition if 1 mol of ethane is reacted in the presence of 5 mol of water at a total pressure of 1 atm.

19.22 Water is added to the feed in the thermal dehydrogenation of ethane in order to prevent the precipitation of solid carbon. For example, carbon could be deposited according to the reaction

$$C_2H_4 \rightarrow 2C + 2H_2,$$

and the carbon could be reacted according to

$$C_2 + H_2O \rightarrow CO + H_2.$$

We should like to investigate just how much water must be added in order to ensure that *no* soot forms in the furnace tubes, but we find the problem to be complicated and to require an orderly approach to the solution. Chemical analysis shows that methane is also formed as an undesired side product and detracts from the desired conversion of ethane to ethylene.

a. You need to formulate an adequate set of equations, presumably including equilibrium relationships and material balances, to determine how many moles of water per mole of ethane in the feed should be included so as to ensure that no carbon will precipitate. State clearly what species are presumed to be present at equilibrium. Solid carbon is presumed to be present but in vanishingly small amount.

b. At the operating temperature of 1520°F, calculate the equilibrium constant

$$K = \exp\left(-\frac{2\mu_C^\circ + 2\mu_{H_2}^* - \mu_{C_2H_4}^*}{RT}\right)$$

for the reaction mentioned first above. What units should be attached to this result?

19.23 In a rocket motor, hydrazine N_2H_4 was reacted with oxygen O_2 at 3500 K and 51 bar. The species listed in the table were found to be present in the mole fractions indicated.

Species	μ_i^*/RT	Final mole fraction
H	−10.021	0.025
H_2	−21.096	0.090
H_2O	−37.986	0.480

Species	μ_i^*/RT	Final mole fraction
N_2	-28.653	0.296
NO	-28.032	?
O	-14.640	0.011
O_2	-30.594	0.023
OH	?	0.059

a. Show that it may reasonably be concluded that the gases were at equilibrium.

b. Calculate, from equilibrium considerations, the expected mole fraction of the species NO and estimate the value of μ_i^*/RT for the species OH.

c. What was the initial fuel to oxygen ratio?

19.24 Calculate the equilibrium composition of a system containing 40 mol of N, 12 mol of O, 30 mol of H, and 1 mol of S. Note that these are given as atoms of the elements, not as molecules. The system at equilibrium is presumed to contain O_2, H_2, H_2O, SO_2, and N_2. The pressure is 2 atm, and the temperature is 1200°F, where we find the following equilibrium constants for the chemical reactions given:

$$H_2 + \frac{1}{2}O_2 \rightarrow H_2O, \quad K_1 = 1.43 \times 10^{11} \text{ atm}^{-1/2},$$

$$3H_2 + SO_2 \rightarrow H_2S + 2H_2O, \quad K_2 = 1.43 \times 10^8 \text{ atm}^{-1},$$

$$2H_2O + 2SO_2 \rightarrow 2H_2S + 3O_2, \quad K_3 = 2.42 \times 10^{-51} \text{ atm}.$$

You are expected to inspect the governing material balances and equilibrium relationships and develop a successive-approximation or trial-and-error method that converges rapidly to the correct answer but which involves the use of only one equation at a time. (You are not supposed to solve the system simultaneously, as we did in the computer problem.) For example, it is immediately evident that the number of moles of nitrogen is $n_{N_2} = 20$.

19.25 Consider the oxidation of sulfur dioxide to sulfur trioxide by reaction with oxygen. Specify the chemical reaction and its corresponding equilibrium constants K and K_y (including units). Calculate the temperature at which the equilibrium constant K becomes equal to unity (in the appropriate units). Use the method of back substitution wherein the difference in $1/T - 1/T_o$ is calculated in the enthalpy-change term. Thus, ln $[K/K(T_o)]$, as well as any terms involving the change in the heat capacity ΔC_p^*, are put on the opposite side of the equation, the correction terms in ΔC_p^* being evaluated at the trial temperature.

19.26 A gas stream is suspected to contain simultaneously NH_3, N_2, H_2, CO, CH_3OH, and H_2CO. Give a complete set of independent reactions that can be used to describe chemical equilibrium in this system. How many independent material balances are there? On which elements?

19.27 The thermal dehydrogenation of ethane is carried out in a constant-temperature and constant-volume reactor to produce ethylene. Initially, 1 mol of ethane and 1 mol of water (an inert) are charged into a vessel at 1150 K and 1 atm total pressure.

Write an expression for the equilibrium constant K for the reaction in terms of the extent of reaction and the final (equilibrium) pressure p_2. Be sure to specify the reaction you are considering. Assume ideal-gas behavior. If the equilibrium pressure is observed to be 1.4585 atm at 1150 K, what is the value of K? What is the equilibrium composition of the gas?

What are the units of K? Does K depend on pressure or on temperature?

19.28 In a waste disposal project, it is desired to oxidize hydrogen cyanide to form less toxic products. Various experiments, under mildly oxidizing conditions, using O_2 as the oxidizing agent, suggested the presence of the following substances: ammonia, cyanogen (NCCN), nitrogen, and nitrous oxide. No analysis was made for compounds not containing nitrogen. Suggest a set of species present in the product stream and determine in an orderly manner a complete set of independent chemical reactions describing interactions among these species.

19.29 Determine the equilibrium constant at 800 K for the gas-phase reaction

$$4HCN + O_2 \rightleftharpoons 2H_2O + 2NCCN.$$

19.30 Calculate the equilibrium composition when 1 mol of sulfur dioxide is reacted with 5 mol of oxygen in the presence of 20 mol of nitrogen (an inert). The product is sulfur trioxide, and the equilibrium constant K is 1 atm$^{-1/2}$ at the operating temperature of 1066.33 K. The pressure is 1 atm, and fugacity-coefficient corrections can be ignored. Use the Newton–Raphson linearization method of successive approximations.

19.31 The equilibrium constant for the hydration of ethylene to form ethanol

$$C_2H_4 + H_2O \rightleftharpoons C_2H_5OH$$

is

$$K = 0.210 \text{ atm}^{-1} \text{ at } 800 \text{ } K.$$

Calculate the equilibrium mole fractions of all three species as functions of the ratio of H_2O to C_2H_4 in the feed (for values of this ratio in the range from 0.3 to 3). Also calculate the conversion of ethylene to ethanol. Make a graph of these results. Assume that the pressure is 2 atm, but that fugacity-coefficient corrections can be ignored.

Does the equilibrium constant K depend on the pressure?

19.32 Determine the equilibrium constant at 800 K for the gas-phase reaction

$$HCN + H_2O \rightleftharpoons NH_3 + CO.$$

19.33 Phosphorus (III) bromide results from the reaction of bromine and red phosphorus

$$P_4 + 6Br_2 \rightarrow 4PBr_3.$$

This reaction occurs in the gas phase. Suppose you have this reaction and some others occurring simultaneously in a reactor at temperature T_o and pressure p. You are using the computer and the multi-variable Newton–Raphson technique to find the equilibrium composition. This is the i'th reaction in the series of reactions. Write the Fortran program segment that sets up the i'th row of the matrix to be solved by "MATINV." Set up the equilibrium equation in terms of moles, not mole fractions or partial pressures. You also know trial values for the numbers of moles:

$$n^*_{PBr_3}, n^*_{Br_2}, \text{and } n^*_{P_4}$$

19.34 For space missions, fuel cells have been an attractive means of making work from the chemical energy stored in gaseous hydrogen and oxygen, and potable water can be produced as a byproduct. Determine the maximum work W_s that can be extracted in the process sketched in Fig. 19.5. You do not need to worry about the details of the electrochemical process because you are to treat a reversible process. Compare your result to the value 145 kJ/mol of O_2 estimated if the hydrogen had been burned and the thermal energy had been converted to work by means of a modified Rankine cycle.

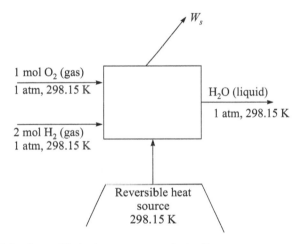

Figure 19.5 Reversible hydrogen–oxygen fuel cell.

19.35 In the new tables of chemical thermodynamic properties (replacing Tables 19.1, 19.2, and 20.1), we wish to change the units for enthalpy and Gibbs function from kcal/mol to kJ/mol and change the pressure unit used for the ideal-gas state from 1 atm to 1 bar. Discuss quantitatively the changes that need to be made for the molar heat capacity, enthalpy, and Gibbs function of formation for the following substances:

a. H_2 ideal-gas state

b. O_2 ideal-gas state

c. C solid

d. CO ideal-gas state

e. CO_2 ideal-gas state

f. H_2O ideal-gas state

g. H_2O liquid

19.36 Calculate the equilibrium constant at 1100 K for the gas-phase hydration of ethylene to form ethanol. Be sure to specify the chemical reaction for which you are determining K. What are the units of K? Does K depend on pressure or on temperature or on composition?

19.37 Initially a system contains 1 mol of water, 0.2 mol of H_2, 1 mol of CO, and 0.3 mol of CO_2. Determine the equilibrium composition at 1100 K if the mixture is equilibrated with respect to the water–gas shift reaction

$$CO + H_2O \rightleftharpoons CO_2 + H_2$$

for which the equilibrium constant is $K = 0.993$ at this temperature.

Notation

a, b, c	constants in Eq. 5.6
$B_{i,j}$	matrix for solution to linearized equations
C	conversion
C_p	heat capacity at constant pressure, J/K
G	Gibbs function, J
ΔG_j^*	standard Gibbs-function change for reaction j, J/mol
G_i	constant terms in matrix equation
H	enthalpy, J
K_j	equilibrium constant for reaction j
K_p	equilibrium ratio of partial pressures
K_y	equilibrium ratio of mole fractions
M_i	symbol for the chemical formula of species i
n	total number of moles, mol
n_i	number of moles of species i, mol
p	pressure, N/m^2
p_i	partial pressure of species i, N/m^2
R	universal gas constant, 8.3143 J/mol-K or 82.06 atm-cm^3/mol-K
S	entropy, J/K
t	time, s
T	thermodynamic temperature, K
T_o	reference temperature, K
V	volume, m^3

y_i mole fraction of species i

λ_i^* secondary reference state quantity for ideal-gas state, atm^{-1}

ν_{ij} stoichiometric coefficient of species i in reaction j

μ_i chemical potential of species i, J/mol

μ_i^* secondary reference state quantity for ideal-gas state, J/mol

ξ_j extent of reaction j

ϕ_i fugacity coefficient of species i

Subscript, superscripts, and special symbols

i species i

o   initial

o   pure component

*   trial solution

*   ideal-gas state

~   per mole

References

1. D. D. Wagman, W. H. Evans, V. B. Parker, I. Halow, S. M. Bailey, and R. H. Schumm. *Selected Values of Chemical Thermodynamic Properties. Tables for the First Thirty-Four Elements in the Standard Order of Arrangement.* NBS Technical Note 270-3. Washington, D.C.: National Bureau of Standards, 1968. [See also NBS Technical Note 270-4 (1969), NBS Technical Note 270-5 (1971), and NBS Circular 500 (1952). The technical notes replace Circular 500 for the elements which they cover.]

2. Frederick D. Rossini, Kenneth S. Pitzer, Raymond L. Arnett, Rita M. Braun, and George C. Pimentel. *Selected Values of Physical and Thermodynamic Properties of Hydrocarbons and Related Compounds.* (American Petroleum Institute Research Project 44.) Pittsburgh: Carnegie Press, 1953.

3. D. R. Stull and H. Prophet. *JANAF Thermochemical Tables,* Second Edition, June, 1971, NSRDS-NBS 37.

Chapter 20

Dilute Solutions

A dilute solution is, in many ways, analogous to a gas mixture, and it is perhaps the next to discuss in order of complexity. In one case, a vacuum provides the medium in which the gaseous components interact; in the other case, the solvent provides the medium in which the solutes interact. A solution approaching infinite dilution (or zero solute concentrations) is similar to a gas in the low-pressure limit or ideal-gas state. Departures from this simple state are observed as the total solution concentration becomes appreciable (or the total pressure becomes appreciable in the case of the gas mixture).

Solutions are more complicated than gas mixtures because there are so many solvents of interest and because the solvent itself has structure and changes with temperature and pressure. Water is the universal solvent, and data for aqueous solutions rival in extent those for gas mixtures. For other solvents, data may be incomplete or absent.

Liquids and solid solutions have one simple feature compared to gas mixtures. The pressure dependence of many physical properties is so slight that it is frequently neglected. Examples 7.3 and 9.1 were designed to illustrate this point in connection with the pressure dependence of the internal energy.

The Newman Lectures on Thermodynamics
John Newman and Vincent Battaglia
Copyright © 2019 Jenny Stanford Publishing Pte. Ltd.
ISBN 978-981-4774-26-0 (Hardcover), 978-1-315-10861-2 (eBook)
www.jennystanford.com

20.1 Mole Ratios

The composition of a solution is frequently expressed in the ratio r_i of the number of moles of solute to the number of moles of solvent (we let the solvent be denoted as component ◦):

$$r_i = \frac{n_i}{n_o}. \tag{20.1}$$

For most chemical engineers, this statement seems to require a lengthy justification. The mole ratio gives emphasis to the special role of the solvent. Since there is one mole ratio for each solute and none for the solvent, we have the correct number of intensive composition variables.

For aqueous solutions, a modified ratio, the *molality m_i*, is used. This is the number of moles of solute per kilogram of solvent and can be expressed as

$$m_i = \frac{n_i}{M_o n_o} = \frac{r_i}{M_o} \tag{20.2}$$

if the molar mass M_o of the solvent is expressed in kg/mol. The molality probably came into use because of its similarity to the molar concentration c_i, expressed in moles of solute per liter of solution. However, the molar concentration is awkward because its value changes with temperature and pressure for a given solution. Furthermore, its evaluation requires a determination of the density, while a solution of a given molality can frequently be prepared accurately by weight. (See also Example 20.2.)

We carry out the development in this book in terms of the molality. In an application requiring mole ratios, m_i can be replaced by r_i, and M_o can be replaced by unity.

In the analysis of processes for gas absorption and liquid extraction, it is convenient to use mole ratios or molalities if one can assume that the solvent is negligibly distributed between the phases. Then the amount of solvent in each stream is constant and can be used as a basis for material balances on the solutes.

When a liquid mixture is treated in terms of mole fractions, no component need be designated as the solvent. This more general situation is developed in Chapter 21.

It should be pointed out again that Gibbs originally used the mass rather than the number of moles in his development of multicomponent thermodynamics. Mass fractions and mass ratios find common use in the industry. Even an international congress cannot change a mass ratio, whereas mole ratios, molalities, and molar concentrations are more vulnerable. On the other hand, moles can make simpler the description of vapor–pressure lowering, freezing-point depression, and boiling-point elevation of liquid solutions.

20.2 Chemical Potentials

For a solution in a given solvent at a given temperature and pressure, let the chemical potential of a solute species be expressed as

$$\mu_i = RT \ln (\lambda_i^\theta m_i \gamma_i) = \mu_i^\theta + RT \ln (m_i \gamma_i). \qquad (20.3)$$

Equation 18.5 is the inspiration for this form. Here γ_i is the *activity coefficient* of component i, taken to approach unity in an infinitely dilute solution (in Chapter 21, the same symbol will unfortunately be used to represent a different activity coefficient):

$$\gamma_i \to 1 \text{ as } \sum_{j \neq 0} m_j \to 0. \qquad (20.4)$$

The quantities λ_i^θ and μ_i^θ depend on temperature and pressure but are independent of composition. The principal composition dependence of the chemical potential is given by the molality m_i in Eq. 20.3, and the activity coefficient γ_i is required to describe any departures from this simple composition dependence. For convenience, γ_i is arbitrarily required to become unity in an infinitely dilute solution, and this value would be a first estimate for any dilute solution.

Thus, λ_i^θ and μ_i^θ are required to describe the principal dependence of the chemical potential on temperature and pressure, although γ_i still has a secondary dependence on T and p. The numerical value of λ_i^θ or μ_i^θ is determined by the condition of Eq. 20.4; for example,

$$\mu_i^\theta = \lim_{m_T \to 0} (\mu_i - RT \ln m_i) \qquad (20.5)$$

where

$$m_T = \sum_{j \neq 0} m_j \tag{20.6}$$

is the total solution molality. The value of μ_i is, of course, set by the *primary reference state* for component i. The steam tables are based on zero enthalpy and entropy for pure liquid water at the temperature and pressure of the triple point. On the other hand, tables of chemical thermodynamics (see Tables 19.1 and 19.2) are based on zero values for the elements in their stable forms at 25°C and 1 atm (or on the ideal-gas state if the element is a gas under these conditions). Table 20.1 provides values of μ_i^θ for aqueous solutions, which are in harmony with the latter convention. Other quantities in the table will be mentioned later.

Equation 20.4 defines a *secondary reference state*, also known as a standard state, in terms of an extrapolation of actual data to infinite dilution. Therefore, λ_i^θ and μ_i^θ are values related to this secondary reference state, and the other factors in Eq. 20.3 describe the composition dependence of μ_i near this secondary reference state. The tables [1] of the National Bureau of Standards say, "The standard state for a solute in aqueous solution is taken as the hypothetical ideal solution of unit molality ... In this state the partial molal enthalpy and heat capacity of the solute are the same as in the infinitely dilute real solution." From this, one must infer that the value tabulated as $\Delta Gf°$ is the same as μ_i^θ given by Eq. 20.5, and "the hypothetical ideal solution of unit molality" means that the same numerical value for m_i in Eq. 20.3 is obtained by setting the quantity $m_i \gamma_i$ equal to 1 mol/kg. With λ_i^θ, we can avoid taking the logarithm of a quantity with dimensions, and λ_i^θ itself has units (kg/mol) reciprocal to those of m_i.

An analogy exists between the quantities λ_i^θ and μ_i^θ on the one hand and λ_i^* and μ_i^* for the ideal-gas state on the other. For the temperature dependence, we have

$$\left(\frac{\partial(\mu_i^\theta / T)}{\partial T} \right)_p = R \left(\frac{\partial \ln \lambda_i^\theta}{\partial T} \right)_p = -\frac{\bar{H}_i^\theta}{T^2}, \tag{20.7}$$

where $\bar{H}_i^\theta$ is the partial molar enthalpy of component i at infinite dilution (see Problem 16.1). Values of $\bar{H}_i^\theta$ at 25°C are also given in

Table 20.1. The temperature derivative of $\bar{H}_i^\theta$ is, in turn, the partial molar heat capacity $\bar{C}_{pi}^\theta$ at infinite dilution. If we had an equation for $\bar{C}_{pi}^\theta$ as a function of temperature, then we could readily calculate $\bar{H}_i^\theta$ and μ_i^θ as a function of temperature, as we did for the ideal-gas state in Chapter 19 and Example 17.4. A lack of data hampers this procedure.

Table 20.1 Quantities characterizing dilute aqueous solutions of undissociated solutes at 25°C and 1 atm.

	$\bar{H}_i^\theta$	μ_i^θ	$\bar{V}_i^\theta$	β	$\lambda_i^\theta/\lambda_i^*$
	kcal/mol	kcal/mol	cm^3/mol	kg/mol	atm-kg/mol
N_2	—	—	32.8	—	—
H_2	−1.0	4.2	25.2	—	1200
O_2	−2.8	3.9	30.4	—	722
Ar	−2.9	3.9	31.7	—	722
CH_4	−21.28	−8.22	37.4	—	735
CO_2	−98.90	−92.26	37.6	−0.015	28.9
HCCH	50.54	51.88	—	—	23.9
NH_2NH_2	8.20	30.6	28.7	—	3.34 × 10^{-6}
CO	−28.91	−28.66	36	—	1047
NH_3	−19.19	−6.35	24	—	0.0171
H_2S	−9.5	−6.66	—	—	9.93
SO_2	−77.194	−71.871	—	—	0.81
C_2H_6	−24.40	−4.09	53.3	—	580
C_2H_4	8.69	19.43	—	—	204
CH_3OH	−58.779	−41.92	38	—	0.0045
CH_3NH_2	−16.77	4.94	—	—	0.01
CH_3COOH	−116.10	−94.78	52	−0.055	1.14 × 10^{-4}
C_2H_5OH	−68.9	−43.44	55.0	−0.063	0.0049
HCN	25.6	28.6	31	—	0.132

Note: The superscript θ denotes the infinite-dilution secondary reference state (1 mol/kg molality unit for μ_i^θ). Primary reference states are based on the elements, in harmony with Tables 19.1 and 19.2. Data on $\bar{H}_i^\theta$ and μ_i^θ are taken from Ref. [1]. Remember that 1 cal = 4.184 J.

For solutions, λ_i^θ and μ_i^θ have a pressure dependence given by

$$\left(\frac{\partial \mu_i^\theta}{\partial p}\right)_T = RT\left(\frac{\partial \ln \lambda_i^\theta}{\partial p}\right)_T = \bar{V}_i^\theta, \qquad (20.8)$$

where $\bar{V}_i^\theta$ is the partial molar volume of component i at infinite dilution. This will be discussed further in Section 20.6.

20.3   Solubility Coefficients

From the data in Table 20.1, let us estimate the solubility of oxygen in water at 25°C. Equation 16.11 requires the chemical potential of oxygen to be the same in two equilibrated phases. For μ_i in the gas phase, use Eq. 18.5, and for μ_i in the aqueous solution, use Eq. 20.3 and thereby obtain

$$\mu_i^\theta + RT \ln (m_i \gamma_i) = \mu_i^* + RT \ln (y_i p \phi_i). \qquad (20.9)$$

To avoid confusion, let p_i equal $y_i p$, the product of the pressure and the mole fraction of oxygen in the vapor phase. Rearrangement gives

$$m_i = p_i \frac{\phi_i}{\gamma_i} \frac{\lambda_i^*}{\lambda_i^\theta}, \qquad (20.10)$$

where

$$\frac{\lambda_i^*}{\lambda_i^\theta} = \exp\left(\frac{\mu_i^* - \mu_i^\theta}{RT}\right) = \exp\left(\frac{-3900}{1.987 \times 298}\right) = 1.384 \times 10^{-3} \frac{\text{mol}}{\text{kg-atm}}$$
$$(20.11)$$

can be designated the *solubility coefficient* and has been evaluated from the data in Tables 20.1 and 19.1.

For a low pressure, we can estimate that the fugacity coefficient ϕ_i is close to unity, and for a very dilute solution, we can estimate that the activity coefficient γ_i of oxygen is also close to unity. For an oxygen partial pressure of 1 atm, we thus estimate the solubility to be $m_i = 1.384 \times 10^{-3}$ mol/kg or $c_i = 1.380 \times 10^{-3}$ mol/L. Davis, Horvath, and Tobias [2] quote the accepted literature value as 1.26×10^{-3} mol/L. The origin of this 10% discrepancy is not known. Data in a compilation like Ref. [1] tend to become dissociated from the original experimental data, and the possibilities of error are many.

A value of μ_i^θ equal to 3.955 kcal/mol for O_2 would be required to reproduce the data.

The reciprocal of the solubility coefficient is one variant of *Henry's constant*. Henry's law expresses the fact that the partial pressure of a component in the vapor in equilibrium with a solution is proportional to the molality m_i at sufficiently high dilutions where γ_i equals unity

$$p_i \rightarrow m_i \frac{\lambda_i^\theta}{\lambda_i^*} \text{ as } m_T \rightarrow 0 .$$

Consequently, we calculate values of $\lambda_i^\theta / \lambda_i^*$ and include them in Table 20.1. A material would be said to be *volatile* relative to water if this ratio is greater than about 5.6×10^{-4} atm-kg/mol, since then the mole ratio (to water) would be larger in the vapor than in the solution. These values of Henry's constants provide the beginning of the thermodynamic background necessary to treat gas absorption.

20.4 Activity Coefficients and Osmotic Coefficient

The solute activity coefficients approach unity as the molality of all solutes approaches zero. As a first approximation to departures from this ideal situation,[1] we may write

$$\ln \gamma_i = 2 \sum_{j \neq 0} \beta_{i,j} m_j , \qquad (20.12)$$

thereby restricting ourselves to linear terms but recognizing that a nonzero molality of one component will generally cause the activity coefficients of all solutes to depart from unity. The coefficients $\beta_{i,j}$ generally depend on temperature and pressure but are independent of composition; we show a little later that they are required to be symmetric,

$$\beta_{i,j} = \beta_{j,i}. \qquad (20.13)$$

---

[1] A solution is properly called *ideal* only when the activity coefficients are unity on a mole fraction basis, not on a molality basis. The molality approaches infinity when the mole fraction of the solvent approaches zero, and the activity coefficient of each solute must approach zero even for an ideal solution.

The corresponding chemical potential of the solvent can be obtained by application of the Gibbs–Duhem Eq. 16.6 at constant temperature and pressure:

$$d\mu_o = -\sum_{i\neq 0} \frac{n_i}{n_o} d\mu_i = -M_o \sum_{i\neq 0} m_i d\mu_i \,. \tag{20.14}$$

Substitution of Eqs. 20.3 and 20.12 yields

$$d\mu_o = -M_o RT \sum_{i\neq 0} \left(dm_i + 2m_i \sum_{j\neq 0} \beta_{i,j} dm_j \right), \tag{20.15}$$

and integration gives

$$\frac{\mu_o - \mu_o^o}{M_o RT} = -m_T - \sum_{i\neq 0}\sum_{j\neq 0} \beta_{i,j} m_i m_j , \tag{20.16}$$

where μ_o^o is the chemical potential of the pure solvent at the given temperature and pressure. This result shows that the chemical potential of the solvent for small m_T depends linearly on the solute molality, while the chemical potential of the solute depends logarithmically on m_i, as in Eq. 20.3. On this basis, we define the *osmotic coefficient* (sometimes called the practical osmotic coefficient) ϕ according to[2]

$$\frac{\mu_o - \mu_o^o}{M_o RT} = -\phi m_T . \tag{20.17}$$

For the activity coefficient variation given by Eq. 20.12, the osmotic coefficient becomes

$$\phi = 1 + \frac{1}{m_T} \sum_{i\neq 0}\sum_{j\neq 0} \beta_{i,j} m_i m_j , \tag{20.18}$$

and ϕ approaches unity as m_T approaches zero.

The *colligative* properties of the solution depend on the variation of the chemical potential of the solvent when solute is added. These properties include the lowering of the partial pressure of the solvent and the resulting elevation of the boiling point of a solution with nonvolatile solutes. Also included are the depression of the freezing point of dilute solutions and the *osmotic pressure*, the pressure

---

[2]Be careful not to confuse the osmotic coefficient with the gas-phase fugacity coefficient ϕ_i.

difference between the solution and pure solvent when the two are equilibrated across a membrane permeable only to the solvent.

We notice that the first term on the right in Eq. 20.16, representing the decrease in the chemical potential of the solvent, is minus the total molality and is independent of the particular chemical nature of the solutes, represented by $\beta_{i,j}$ in the next term. Consequently, the colligative properties of very dilute solutions depend only on the number of moles of solute and can be used to determine the molar mass of a polymer in solution and to determine whether a solute dissociates when it is dissolved.

20.5   Thermodynamic Consistency

We have emphasized even in Chapter 1 the interrelationships prevalent in thermodynamics. These can be used to advantage to calculate a result that has not been measured experimentally. They can also be used to test the accuracy of experimental data that are interdependent and to test the validity of a theory or method of calculation. In Eq. 20.12, we propose a composition dependence for the activity coefficients. How can we be sure that these are internally consistent?

As indicated at the beginning of Chapter 16, Gibbs showed that a knowledge of G as a function of T, p, n_j permits any thermodynamic quantity to be obtained in a consistent manner. Consequently, let us express the total Gibbs function G for the solution according to Eq. 14.12. Substitution of Eqs. 20.3 and 20.12 for the chemical potential of the solutes and Eq. 20.16 for that of the solvent leads to

$$\frac{G}{RT} = \frac{n_o \mu_o^o}{RT} + \sum_{i \neq 0} n_i \left[\ln (m_i \lambda_i^\theta) - 1 \right] + \sum_{i \neq 0} \sum_{j \neq 0} \beta_{i,j} n_i m_j . \quad (20.19)$$

Here the parameters μ_o^o, λ_j^θ, and $\beta_{i,j}$ depend on temperature and pressure. Now it can be verified that Eqs. 20.12 and 20.16 or 20.18 are consistent by obtaining them from the Gibbs function of Eq. 20.19 by differentiation according to Eq. 14.14.

Since $n_i m_j = n_j m_i$, the coefficients $\beta_{i,j}$ and $\beta_{j,i}$ occur as coefficients of identical factors, and they cannot be determined separately. Equation 20.13 sets them equal to each other. Otherwise, the sum of $\beta_{i,j}$ and $\beta_{i,j}$ would occur in place of $2\beta_{i,j}$ when Eq. 20.12 is obtained from Eq. 20.19 by differentiation.

Some say that thermodynamic consistency is assured if one can integrate the Gibbs–Duhem equation, as we did to obtain Eq. 20.16, without finding any contradiction. Thus, the right side of Eq. 20.15 must be a perfect differential. Otherwise different values for the variation of μ_o could be obtained by integrating along different paths from one composition to another.

Another approach to thermodynamic consistency is based on Eq. 16.4. If neither species i nor species k is the solvent, this can be rewritten as

$$\left(\frac{\partial \mu_i}{\partial m_k}\right)_{\substack{T,p,m_j \\ j \neq k}} = \left(\frac{\partial \mu_k}{\partial m_i}\right)_{\substack{T,p,m_j \\ j \neq i}}, \tag{20.20}$$

and substitution of Eq. 20.3 yields

$$\left(\frac{\partial \ln \gamma_i}{\partial m_k}\right)_{\substack{T,p,m_j \\ j \neq k}} = \left(\frac{\partial \ln \gamma_k}{\partial m_i}\right)_{\substack{T,p,m_j \\ j \neq i}}. \tag{20.21}$$

It can be seen that Eq. 20.12 is self-consistent on this basis if the β coefficients are symmetric according to Eq. 20.13. This last method allows one to discover an inconsistency among data or expressions for the chemical potentials of as few as two components in the absence of information on the remaining components.

20.6 Pressure Dependence

Equation 20.8 shows that the partial molar volume holds the key to the pressure dependence of the solubility coefficient. At 25°C and for a partial molar volume of 20 cm³/mol, we can calculate that Henry's constant $\lambda_i^\theta / \lambda_i^*$ increases by 1% for each increase in total pressure of 12.2 atm. An increase in 4.9 atm is required for $\overline{V}_i^\theta = 50$ cm³/mol. Correspondingly, the solubility coefficient decreases by about 1%. For moderate pressure variations, changes in the solubility coefficient can usually be ignored.

The partial molar volume itself can be obtained from density data for the solution. Figure 20.1 shows $\hat{V}$ plotted against the mass fraction of ammonia. A straight line tangent to this curve takes the simple form

$$\hat{V} = \frac{\overline{V}_0}{M_0} + \left(\frac{\overline{V}_1}{M_1} - \frac{\overline{V}_0}{M_0} \right) \omega_1 , \qquad (20.22)$$

where 1 refers to the solute, ω_1 is the mass fraction, and the partial molar volumes apply at the tangent point. Thus, the slope and the intercept of a tangent line yield the partial molar volumes of the two components of a binary solution. The curve in Fig. 20.1 indicates that the partial molar volume of ammonia decreases with an increase in concentration, while that of water increases.

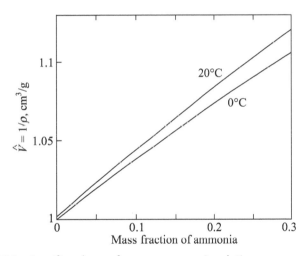

Figure 20.1 Specific volume of aqueous ammonia solutions.

The volumetric data in Fig. 20.2 for aqueous methanol solutions show a point of inflection. Consequently, the partial molar volume of methanol at 20°C begins at about 37.9 cm³/mol, decreases to a minimum of about 36.4 cm³/mol, and then rises to 40.5 cm³/mol for pure methanol.

A plot of molar volume versus mole fraction yields an even simpler equation for a tangent line:

$$\tilde{V} = \overline{V}_0 + (\overline{V}_1 - \overline{V}_0) x_1 . \qquad (20.23)$$

For many solutions, the density is tabulated as a function of the concentration (moles per unit volume) of the solute. The partial molar volume of the solvent is then given by

$$\bar{V}_0 = \frac{M_0}{\rho - c_1 \dfrac{d\rho}{dc_1}}, \tag{20.24}$$

while that of the solute is

$$\bar{V}_1 = \frac{M_1 - \dfrac{d\rho}{dc_1}}{\rho - c_1 \dfrac{d\rho}{dc_1}}. \tag{20.25}$$

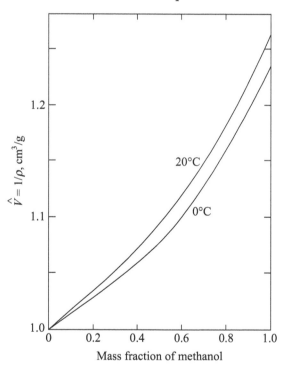

Figure 20.2 Specific volume of aqueous methanol solutions.

Equation 14.10 is a combination of these two formulas. Multisolute formulas are also available, see Appendix A in Ref. [3].

A dilatometer is a device for measuring the partial molar volume of a sparingly soluble gas. It involves the accurate measurement of the volume change when a given amount of gas is dissolved in a fixed amount of solvent or solution. This is superior to the separate determination of the densities of the solution and the pure solvent. In fact, the procedure with the dilatometer is essentially identical to that implied by the basic defining Eq. 14.7 for the partial molar volume. That equation involves the volume change when a differential number of moles of solute is added to a fixed number of moles of solvent.

20.7 Mole-Fraction Basis

Most chemical-engineering applications of thermodynamics use mole fractions as the means of expressing the composition, and this usually includes dilute, nonaqueous solutions. The physical properties used with mole fractions are different from but related to those used with molalities. We explore the connection here.

We now express the chemical potential of a solute as

$$\mu_i = RT \ln (\Lambda_i^\theta x_i \Gamma_i),$$ (20.26)

where Γ_i is the activity coefficient normalized to approach unity at infinite dilution of all solutes and Λ_i^θ is a quantity related to the secondary reference state thus defined. We use capital letters for these quantities to avoid confusion with γ_i and λ_i^θ. Comparison with Eq. 20.3 shows that

$$\Lambda_i^\theta x_i \Gamma_i = \lambda_i^\theta m_i \gamma_i.$$ (20.27)

This reference state of infinite dilution is essentially the same as that used in Section 20.2 for dilute solutions on a molality basis. The connection between the mole fraction and the molality of a solute is

$$m_i = \frac{x_i}{M_o x_o} = \frac{x_i/M_o}{1 - \displaystyle\sum_{j \neq 0} x_i}$$ (20.28)

or

$$x_i = \frac{m_i}{1/M_o + m_T}.$$ (20.29)

Near infinite dilution, both γ_i and Γ_i approach unity, and x_i approaches $M_o m_i$. From Eq. 20.27, we can immediately establish the connection between Λ_i^θ and λ_i^θ:

$$\lambda_i^\theta = M_o \Lambda_i^\theta.$$ (20.30)

In analogy to Eq. 20.19, let us express the Gibbs function G of the solution according to

$$\frac{G}{RT} = \frac{n_o \mu_o^o}{RT} + \sum_{i \neq 0} n_i \ln \Lambda_i^\theta + \sum_i n_i \ln x_i$$

$$- \sum_{i \neq 0} \sum_{j \neq 0} \frac{A_{i,j}}{RT} n_i x_j . \tag{20.31}$$

The first two terms on the right are related to the secondary reference states for the components; the next term is the ideal Gibbs free energy of mixing, identical to that for ideal-gas mixtures (see Eq. 17.21). The last term represents nonideal solution behavior by means of Margules-type parameters (see Chapter 21). All possible quadratic terms are included here, since the mole fraction x_0 of the solvent is not an independent composition variable.

If we now replace the mole fractions by molalities according to Eq. 20.29, noting that $x_o = 1 - \sum_{j \neq 0} x_j$, and make expansions of $\ln x_o$ appropriate to dilute solutions, we obtain

$$\frac{G}{RT} = \frac{n_o \mu_o^o}{RT} + \sum_{i \neq 0} n_i \left[\ln \left(M_o \Lambda_i^\theta m_i \right) - 1 \right]$$

$$- M_o \sum_{i \neq 0} \sum_{j \neq 0} \left(\frac{A_{i,j}}{RT} + \frac{1}{2} \right) n_i m_j . \tag{20.32}$$

Comparison with Eq. 20.19 leads again to Eq. 20.30 for the relationship between λ_i^θ and Λ_i^θ and requires that

$$\beta_{i,j} = -M_o \left(\frac{A_{i,j}}{RT} + \frac{1}{2} \right), \tag{20.33}$$

thereby assuring that $A_{i,j}$ is also symmetric.

Equation 20.31 is not equivalent to Eq. 20.32 and hence not equivalent to Eq. 20.19, since terms of order $n_i m_j^2$ have been neglected. In fact, we can add terms to both Eqs. 20.19 and 20.31 to account for solution nonidealities not expressed adequately by the terms in $\beta_{i,j}$ or $A_{i,j}$. Equation 20.33 is necessary to assure agreement of the two expansions for dilute solutions. The two truncated series will give divergent results in concentrated solutions. The situation is analogous to the use of the two virial Eqs. 3.24 and 3.25 discussed in Section 3.8.

Now, differentiation of Eq. 20.31 according to Eq. 14.14 leads to the activity coefficient of a solute

$$\ln \Gamma_i = -2\sum_{j\neq0} \frac{A_{i,j}}{RT} x_j + \sum_{j\neq0}\sum_{k\neq0} \frac{A_{j,k}}{RT} x_j x_k,$$ (20.34)

while the chemical potential of the solvent is given by

$$\frac{\mu_o - \mu_o^o}{RT} = \ln x_o + \sum_{i\neq0}\sum_{j\neq0} \frac{A_{i,j}}{RT} x_i x_j.$$ (20.35)

20.8 Electrolytic Solutions

The preceding sections deal with nonelectrolytes. Electrolytic solutions contain ions in addition to a solvent. Ions have very strong interactions which lead to important consequences in their thermodynamics. First, the resulting solutions are electrically neutral to a very good approximation because it takes a great deal of energy to separate charges. This means that a solution of Cl⁻ and Na⁺ ions has three species but only two independent components, H_2O and NaCl. Second, the interionic forces cause ions to be distributed preferentially around an ion of opposite charge. This means that in dilute solutions the activity coefficients vary linearly with the square root of concentration, instead of linearly with the concentration, as is the case with nonelectrolytes (see Eq. 20.12). This anomalous behavior caused intensive research in the early part of the twentieth century. Debye and Hückel [5] clarified the behavior in 1923 by giving quantitative treatment to the distribution of ions around a central ion and by showing how this distribution contributes to the Helmholtz and Gibbs energies and the resulting activity coefficients.

Electrolytic solutions generate an extensive body of knowledge. We summarize this here but refer the reader to Ref. [3] for details.

We think that we learned everything about electricity and magnetism in physics classes. However, this generally dealt with charged objects in a vacuum or in a low-pressure gas. In condensed phases ions interact at very short distances, and Coulomb's law is merely an asymptotic form of the intermolecular force for long distances. Consequently, it is a challenge even to define the electric potential in such a medium. The practical way is to use a reference electrode, reversible to one of the ions in the solution, and this remains also the simplest theoretical way to approach the subject. While an outer potential can be defined at an air/solution interface, its measurement can constitute a PhD dissertation, and the result of all that effort really does not have a consequence in thermodynamics. The value of the potential difference of an electrochemical cell can be predicted without using the result for the outer potential.

As a consequence of the strong electric forces, the electrochemical potential μ_i (as the chemical potential of an ion is called) depends on the electrical state of the solution, and this state in turn depends on the electrical environment outside the solution. Activity coefficients are then reported only for electrically neutral combinations of ions, such as γ_{NaCl} instead of activity coefficients of the individual Cl⁻ and Na⁺ ions. In this way, the electrical state cancels between the two ions. Problems 18.3 and 18.4 illustrate some of the techniques for treating the behavior of dilute solutions, where a solvent can be identified. Such methods can be extended to treat electrolytic solutions.

The thermodynamic theory of electrochemistry is fascinating in its own right. One can facilitate and control electrochemical reactions by applying an electric potential between two electrodes. The study of electrochemical systems includes thermodynamics, chemical kinetics, transport phenomena, and materials science. In addition, you have the electric potential to ponder and to use to effect transport and reaction. Then, there are many practical things which can be done with electrical systems. Storage of energy in batteries has always been important, and new inventions are always possible since the chemicals are not specified, but only the overall storage objective. Some important and some esoteric chemicals are produced electrochemically, such as aluminum and chlorine and caustic soda (NaOH) and bleach. Corrosion is frequently an electrochemical process, and preventing such corrosion is important economically. Surface finishing and electronics profit by electrochemical processes. Electrochemistry is important even in biology since organisms rely on ionic solutions and nerve action is related to electrical impulses. Also, electrochemical systems provide a convenient and accurate means to make transport measurements.

Example 20.1 Davis, Horvath, and Tobias [2] express the solubility of oxygen in aqueous KOH solutions as

$$\log\left(\frac{c_{O_2}}{c_{O_2}^o}\right) = -Bc_{KOH},$$

where the oxygen concentrations refer to a partial pressure of 1 atm and B is found to have the value 0.1746 L/mol. $c_{O_2}^o$ is the solubility in pure water, cited to be 1.26×10^{-3} mol/L in Section 20.3. Show how the value of B can be converted into a value of $\beta_{O_2, KOH}$ that appears in Eq. 20.12.

Solution: Assume that the oxygen concentration is so low that it can be neglected in the volumetric relationships. Also approximate $M_o / \overline{V}_o$ by ρ^o, the density of the pure solvent. The equations

$$m_i = \frac{c_i}{c_o M_o}$$

and

$$c_o \overline{V}_o + c_{KOH} \overline{V}_{KOH} = 1$$

(see Eqs. 20.2 and 14.11) can be combined to yield a conversion from concentration to molality:

$$c_i = \frac{\rho^o m_i}{1 + \rho^o m_{KOH} \overline{V}_{KOH}}.$$

The solubility data can, therefore, be re-expressed as

$$\ln\left(\frac{m_{O_2}}{m_{O_2}^o}\right) = -\frac{2.303 \, B\rho^o m_{KOH}}{1 + \rho^o m_{KOH} \overline{V}_{KOH}} + \ln\left(1 + \rho^o m_{KOH} \overline{V}_{KOH}\right).$$

On the other hand, Eq. 20.10 can be written as

$$\ln\left(\frac{m_{O_2}}{m_{O_2}^o}\right) = -\ln \gamma_{O_2} = -2\beta_{O_2, KOH} m_{KOH},$$

where we interpret $p_i \phi_i \lambda_i^* / \lambda_i^\theta$ as $m_{O_2}^o$, the solubility in the absence of KOH, and we use Eq. 20.12 to express the activity coefficient, again neglecting a term that should be small for the small oxygen concentrations encountered here.

Comparison of the last two equations shows that to the first order in m_{KOH}, we can make the association

$$2\beta_{O_2, KOH} = \rho^o (2.303\,B - \overline{V}^{\,\theta}_{KOH}).$$

Substitution of numerical values, including 10.8 cm^3/mol for the partial molar volume of KOH at infinite dilution, gives

$$\beta_{O_2, KOH} = \frac{0.99707}{2}\left(2.303 \times 0.1746 - \frac{10.8}{1000}\right) = 0.195 \frac{kg}{mol}.$$

Example 20.2 When the concentration is given by means of molar concentrations, the chemical potential of a solute is written as

$$\mu_i = RT \ln (a_i^\theta c_i f_i),$$

where f_i is an activity coefficient and a_i^θ characterizes the secondary reference state, again taken to be infinite dilution. The truncated virial expansion for f_i is

$$\ln f_i = 2 \sum_{j \neq 0} \beta'_{i,j} c_j.$$

Discuss the thermodynamic consistency of this formulation and relate the physical properties to those on a molality basis.

Solution: Equating the two expressions for the chemical potential yields

$$\lambda_i^\theta m_i \gamma_i = a_i^\theta c_i f_i.$$

Since $m_i = c_i / M_o c_o$ and f_i and γ_i approach unity, we have

$$\lambda_i^\theta = M_o c_o a_i^\theta = \rho^o a_i^\theta,$$

where ρ^o is the density of the pure solvent. (For λ_i^θ in kg/mol and a_i^θ in L/mol, ρ^o has the units g/cm^3.)

Now, the relationship of the activity coefficients is

$$f_i = \frac{\rho^o}{M_o c_o} \gamma_i$$

or

$$2 \sum_{j \neq 0} \beta'_{i,j} c_j = 2 \sum_{j \neq 0} \beta_{i,j} m_j - \ln \left(\frac{M_o c_o}{\rho^o}\right).$$

Expansion for small concentrations yields

$$\ln \left(\frac{M_o c_o}{\rho^o}\right) = \ln \left(\frac{M_o}{\overline{V}_o \rho^o}\right) + \ln \left(1 - \sum_{j \neq 0} c_j \overline{V}_j\right) = -\sum_{j \neq 0} c_j \overline{V}_j + O(c^2)$$

since $M_0/\bar{V}_0$ is equal to ρ^0 plus an error proportional to c^2. Approximating now m_j by c_j/ρ^0 plus an error proportional to c^2, we obtain

$$\sum_{j\neq0}\left(\beta'_{i,j} - \frac{\beta_{i,j}}{\rho^0} - \frac{\bar{V}_j}{2}\right)c_j = O(c^2).$$

For this to be true for arbitrary dilute solutions requires that

$$\beta'_{i,j} = \frac{\beta_{i,j}}{\rho^0} + \frac{\bar{V}^\theta_j}{2},$$

a result which is not symmetric.

Let us now examine the thermodynamic consistency by means of Eq. 16.4, where $i \neq k$ and neither i nor k is the solvent. Since $c_i = n_i/V$,

$$\frac{1}{RT}\left(\frac{\partial\mu_i}{\partial n_k}\right)_{T,p,n_j \atop j\neq k} = \frac{1}{V}\left[-\bar{V}_k + 2\beta'_{i,k} - 2\bar{V}_k\sum_{j\neq0}\beta'_{i,j}c_j\right].$$

This is not exactly equal to the quantity obtained by interchanging the subscripts i and k, even if we relate $\beta'_{i,j}$ to $\beta_{i,j}$ according to the formula obtained above. It is particularly interesting to note that the formulation for molar concentrations is thermodynamically inconsistent if we set $f_i = 1$, a situation we might be tempted to call "ideal."

Example 20.3 From data on the vapor pressure of aqueous sucrose solutions at 25°C, determine the activity coefficient of sucrose.

Solution: Since sucrose is not volatile and the vapor phase is nearly ideal, the vapor pressure is a measure of the chemical potential of the solvent according to

$$\mu_0 = RT\ln(\lambda^*_0 p).$$

The results have been expressed in Fig. 20.3 in terms of the osmotic coefficient of the solvent in the solution:

$$\frac{\mu_0 - \mu^0_0}{RT} = \ln\left(\frac{p}{p^0}\right) = -\phi M_0 m,$$

where m denotes the molality of sucrose in the solution and p^0 is the vapor pressure of the pure solvent.

Once careful measurements of the vapor pressure have been made for one solute as a function of composition, other nonvolatile solutes can be studied by so-called *isopiestic* measurements; their solution is equilibrated through the vapor phase with a solution of the reference solute. Analysis of the molalities of the equilibrated solutions yields the vapor pressure of the solution being studied. Because extremely accurate temperature control and absolute pressure measurement are not necessary, solutions of many solutes have been investigated by the isopiestic method.

The curve in Fig. 20.3 is approximately, but not exactly, a straight line. Equation 20.18 reduces to $\phi - 1 = \beta m$ for a single solute, and a value $\beta = 0.09$ kg/mol might give an adequate representation of the data. The activity coefficient of sucrose would then follow immediately from Eq. 20.12: $\ln \gamma = 2\beta m$.

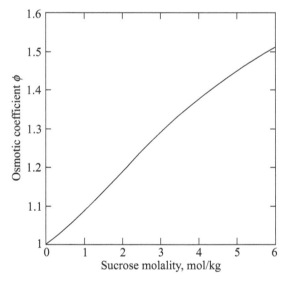

Figure 20.3 Osmotic coefficient of aqueous sucrose solutions at 25°C.

A more rigorous treatment of the data is possible and desirable for the maximum utilization of accurate data. We choose a numerical integration of the Gibbs–Duhem Eq. 16.6. Thus,

$$d\mu_o = -M_o m \, d\mu_1.$$

Substitution of Eq. 20.17 for the solvent and Eq. 20.3 for the solute gives the differential relation between the activity coefficient and the osmotic coefficient:

$$d(\phi \, m) = dm + m \, d \ln \gamma$$

or

$$d(\phi m) - dm = d[(\phi - 1)m] = m \, d \ln \gamma.$$

Integration gives

$$\ln \gamma = \int \frac{1}{m} d \, [(\phi - 1)m] = \frac{(\phi - 1)m}{m} + \int \frac{(\phi - 1)m}{m^2} dm \,,$$

where integration by parts has also been used. The integration constant can be evaluated since γ and ϕ both become unity at infinite dilution and the integrand appears to be well behaved as m approaches zero. Now we have

$$\ln \gamma = \phi - 1 + \int_{0}^{m} \frac{\phi - 1}{m} dm \,.$$

Figure 20.4 plots $(\phi - 1)/m$ versus m, based on the data in Fig. 20.3. We note first that this graph gives an expanded view of the nonlinear behavior depicted in Fig. 20.3. The value of β should be taken to be about 0.078 kg/mol in order to represent correctly the slope of Fig. 20.3 at infinite dilution; more terms should, strictly, be added to Eqs. 20.12, 20.18, and 20.19 to account for the nonlinear behavior. The activity coefficient is now calculated from the area below the curve of Fig. 20.4. This result is shown in Fig. 20.5.

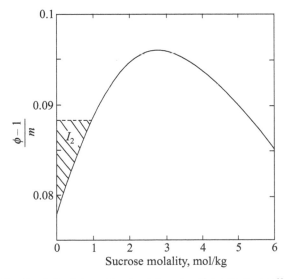

Figure 20.4 Graph for integration to determine the activity coefficient from the osmotic coefficient.

The curve in Fig. 20.5 is certainly very similar to that in Fig. 20.3. To bring this out more clearly, we can integrate by parts again to obtain

$$\ln \gamma = 2(\phi - 1) - \int_{m=0}^{m=m} m\, d\left(\frac{\phi - 1}{m}\right).$$

To a first approximation, $\ln \gamma$ is twice the departure of the osmotic coefficient from unity. The integral can now be interpreted as that area I_2 (drawn for $m = 1$ mol/kg) between the curve in Fig. 20.4 and the ordinate axis; this area is apparently smaller than the area below the curve and would be zero if a straight line satisfactorily represented the curve in Fig. 20.3.

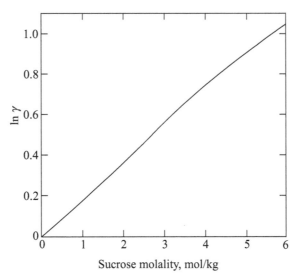

Figure 20.5 Activity coefficient on a logarithmic scale, of aqueous sucrose solutions at 25°C.

Problems

20.1 Show that the pressure dependence of the activity coefficient is given by

$$\left(\frac{\partial \ln \gamma_i}{\partial p}\right)_{T,m_j} = \frac{\bar{V}_i - \bar{V}_i^\theta}{RT}$$

and the temperature dependence is

$$\left(\frac{\partial \ln \gamma_i}{\partial T}\right)_{p,m_j} = \frac{\bar{H}_i^\theta - \bar{H}_i}{RT^2}.$$

Show further that the temperature dependence of β follows from

$$\left(\frac{\partial \beta_{i,j}}{\partial T}\right)_P = -\frac{1}{2RT^2}\frac{\partial \bar{H}_i}{\partial m_j}\bigg|_{m_T=0}.$$

20.2 Estimate Henry's constant for absorption of CO_2 into water at 50°C based on the information in Tables 19.1 and 20.1. You can compare your result with the values reported by Gibbs and van Ness [4]:

T (°C)	$\lambda_i^\theta / M_o \lambda_i^*$ (atm)	$A/RT = -(0.5 + \beta/M_o)$
0	670	−2.420
10	1000	−2.520
15	1233	−1.748
20	1432	−0.700
25	1633	0.315
35	2098	1.478
50	2904	1.948

20.3 Show that Eq. 20.34 is thermodynamically inconsistent if the quadratic terms in the mole fractions are dropped, an approximation that might seem appropriate for dilute solutions.

20.4 a. Distinguish between the terms primary reference state and secondary reference state.

   b. Would the primary reference state or the secondary reference state be expected to be characterized by a series of values (of enthalpy, entropy, and Gibbs function) showing the dependence on temperature (and possibly on pressure)? What sort of values would characterize the other reference state?

20.5 The solubility ratio p_i/m_i for carbon dioxide dissolved in water at 25°C is 29.61 atm-kg/mol for a CO_2 partial pressure of 1 atm, but p_i/m_i has a value of 36.83 atm-kg/mol for a CO_2 partial pressure of 36 atm. What portions of this change should be ascribed separately to changes in the fugacity coefficient ϕ_i, the activity coefficient γ_i, and the secondary reference state quantity λ_i^θ? (The second virial coefficient for CO_2 is -125 cm^3/mol, although you should have been able to estimate it from Fig. 3.5 or Eq. 3.23 and Table 3.1.)

20.6 Sulfur dioxide is to be absorbed in water. It is noted that Henry's constant is 0.81 atm-kg/mol at 25°C. Evaluate the suggestion that chilled water be used in the absorption process by estimating Henry's constant at 15°C.

20.7 Calculate the *osmotic pressure* for a sucrose solution at a molality of 1 mol/kg and a temperature of 25°C. This solution is equilibrated with pure water at 1 atm through a membrane permeable to water but not to sucrose. The lowering of the chemical potential of water due to the presence of the sucrose must be compensated by an increase in the pressure. What pressure is required?

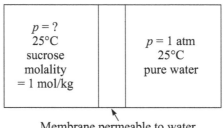

Membrane permeable to water

20.8 a. What are examples of colligative properties? What common feature causes them to be listed under this heading?

b. What are the advantages of using mole ratios (or molalities) as composition units?

c. Why is it helpful to use an unsymmetric convention for secondary reference states when dealing with dilute solutions?

20.9 For a gas-absorption process, let us define the separation factor as the ratio of mole fraction in the gas to the molality in the liquid for one component divided by a similar ratio for the second component. Calculate this separation factor for CO_2 relative to SO_2 for gas absorption in water. Which component goes preferentially into the liquid?

20.10 Determine the mutual solubility of water in carbon tetrachloride and of carbon tetrachloride in water at 25°C. At this temperature, the following data are given:

$$\mu^{o}_{H_2O} = -237.129 \text{ kJ/mol}$$

$$\mu^{\square}_{H_2O} = -219.485 \text{ kJ/mol}$$

$$\mu^{o}_{CCl_4} = -65.21 \text{ kJ/mol}$$

$$\mu^{\theta}_{CCl_4} = -52.131 \text{ kJ/mol} .$$

Data for H_2O in CCl_4 use a mole ratio basis (r_i) and are denoted by a superscript $\square$. Data for CCl_4 in H_2O use a molality basis (m_i) and are denoted by a superscript θ.

20.11 A number of substances are contacted in a vessel maintained at 25°C and 10 atm. In one case, we have

H_2O	1 kg
N_2	2 mol
CO_2	3 mol
H_2S	0.1 mol
SO_2	0.2 mol
H_2	5 mol
CO	0.1 mol.

The general task is to devise a computer program, based on dilute-solution theory, for calculating the composition of both a vapor and a liquid phase. Your part in this task is to formulate governing equilibrium relations, material balances, and physical properties. You are supposed to make sure that this mathematical description is complete but not redundant. It is presumed that the principal unknowns are to be the numbers of moles of each species in each phase, and needed physical-property values should be related clearly to these variables.

20.12 Suppose that aqueous sucrose solutions are being used as a standard of water activity in isopiestic measurements. That is, the osmotic coefficient of these solutions is known in detail as a function of molality.

Now measurements are carried out on solutions of species A to yield the following pairs, which are in isopiestic equilibrium:

Molality of sucrose	Molality of species A
0.1	0.1015
0.5	0.540
1	1.172
2	2.846
3	5.895.

Determine the composition dependence of the activity coefficient of species A. You may assume that sucrose solutions are adequately described by a single β value of 0.09 kg/mol.

20.13 The solubility of CO_2 in CH_3OH in cm^3 (of gas at 0°C and 1 atm) per cm^3 of liquid is

t (°C)	Solubility
15	4.366
20	3.918
25	3.515.

Calculate the heat of solution.

20.14 The specific enthalpy of mixtures of sulfuric acid (component 1) and water (component 2) at 25°C can be expressed approximately as

$$\widehat{H}\,(\text{Btu/lb}) = 44.5\,\omega_2 - 171.5\,\omega_1\omega_2 - 687\omega_1\omega_2^2.$$

where ω_1 is the mass fraction of sulfuric acid and ω_2 is the mass fraction of water. From this expression, we could obtain the partial molar enthalpy for both sulfuric acid ($\bar{H}_1$) and water ($\bar{H}_2$) as follows:

$$\frac{\bar{H}_1}{M_1} = -171.5\,\omega_2^2 - 687\omega_2^2(\omega_2 - \omega_1)$$

and

$$\frac{\bar{H}_2}{M_2} = 44.5 - 171.5\,\omega_1^2 - 2(687)\,\omega_1^2\omega_2.$$

For a dilute-solution treatment of these mixtures, we write

$$\mu_1 = \mu_1^\theta + \nu RT \ln (m\gamma_\pm)$$

and

$$\mu_2 = \mu_2^\theta - \nu RT \, M_2 m \, \phi \,,$$

where ν is taken to be 3 because of the dissociation of sulfuric acid, where m is the molality of sulfuric acid ($m = \omega_1/\omega_2 M_1$), $\gamma_\pm$ is called the mean molal activity coefficient, and ϕ is called the osmotic coefficient.

If the specific enthalpy expression can be taken to be independent of temperature, develop equations for calculating $\gamma_\pm$ and ϕ at a temperature T from the data available at a reference temperature of T_0.

References

1. D. D. Wagman, W. H. Evans, V. B. Parker, I. Halow, S. M. Bailey, and R. H. Schumm. *Selected Values of Chemical Thermodynamic Properties. Tables of the First Thirty-Four Element in the Standard Order of Arrangement.* NBS Technical Note 270-3, Washington, D. C.: National Bureau of Standards, 1968.

2. R. E. Davis, G. L. Horvath, and C. W. Tobias. "The solubility and diffusion coefficient of oxygen in potassium hydroxide solutions." *Electrochimica Acta*, **12**, 287–297 (1967).

3. John Newman and Karen E. Thomas-Alyea. *Electrochemical Systems.* Hoboken, New Jersey: John Wiley & Sons, Inc., 2004, Appendix A.

4. R. E. Gibbs and H. C. van Ness. "Solubility of gases in liquids in relation to the partial molar volumes of the solute. Carbon dioxide–water." *Industrial and Engineering Chemistry Fundamentals*, **10**, 312–315 (1971).

5. P. Debye and E. Hückel, "Zur Theorie der Elektrolyte," *Physikalische Zeitschrift*, **24**, 185–206 (1923).

Chapter 21

Liquid Mixtures

In gas-phase mixtures covering the whole composition range, a virial expansion in the pressure provided a useful approach. In dilute solutions, virial expansions in the solute concentrations were possible. Now we wish to consider the whole range of compositions for a liquid phase, and again we seek a simple model—the ideal solution—as a point of departure. Here a few simple equations describe the mixture, based on a few data on the pure components. The ideal solution provides an adequate description of distillation of petroleum fractions, where the components are chemically similar to each other.

However, a number of interesting phenomena of technological importance involve departures from ideal behavior. It is then expedient to consider specific systems for which experimental data are available. These can include examples in binary distillation and azeotropes, liquid extraction and phase separation, and solid–liquid equilibria. General thermodynamic principles find application in these specific examples, but a general correlation of the thermodynamic properties of multicomponent systems inevitably leads to the loss of some experimental detail. The readers will progress not only by seeking general methods for the estimation of thermodynamic properties from incomplete data but also by studying in detail these specific systems. Different simplifications

The Newman Lectures on Thermodynamics
John Newman and Vincent Battaglia
Copyright © 2019 Jenny Stanford Publishing Pte. Ltd.
ISBN 978-981-4774-26-0 (Hardcover), 978-1-315-10861-2 (eBook)
www.jennystanford.com

apply to different systems, and one can take no advantage of these in a perfectly general approach.

In this chapter, we define the activity coefficient γ_i on a mole-fraction basis. Thereby, we write for the chemical potential of component i in the mixture

$$\mu_i = \mu_i^o + RT \ln(x_i \gamma_i), \tag{21.1}$$

where μ_i^o is the chemical potential of pure component i at the temperature and pressure of the solution, where component i is assumed to be a liquid. The *secondary reference state* for component i is specified by this definition of μ_i^o. For an ideal solution, $\gamma_i = 1$, and the properties of the solution depend essentially on $\mu_i^o(T, p)$. In general, the activity coefficient depends on composition and describes departures from ideal behavior.[1] It must also be expected to be a weak function of temperature and pressure, although the principal dependence on these quantities is contained in μ_i^o.

It takes relatively little space to describe μ_i^o, since it is independent of composition and is closely related to the vapor pressure of pure component i. Descriptions of γ_i, of nonideal behavior, can be lengthy. Nevertheless, we should never lose sight of the simplicity of the ideal-solution approximation and the fact that γ_i represents only a correction to the temperature, pressure, and composition dependence of μ_i represented by the other quantities in Eq. 21.1.

21.1 Vapor as a Meter for Chemical Potentials in Liquid

It may be useful to begin by considering how we might measure the chemical potential in a liquid mixture. The general condition of phase equilibrium, developed in Eq. 16.11, reads

$$\mu_i^{\text{liq}} = \mu_i^{\text{vap}}. \tag{21.2}$$

Since we have already studied extensively the chemical potentials of a vapor phase in Chapters 17 and 18, this relation gives ready access to values of μ_i^{liq}. Furthermore, some of the most important

---

[1]The same symbol was used in Chapter 20 to represent a different activity coefficient—one based on molalities as the composition variables and based on a secondary reference state at infinite dilution of all the solutes.

applications of multicomponent thermodynamics involve vapor–liquid equilibria, and in this way we can also get some insight into this area.

In the vapor phase, we expressed the chemical potential of species i as (see Eq. 18.5)

$$\mu_i = \mu_i^*(T) + RT \ln (py_i\phi_i).$$
(21.3)

At low pressures, the fugacity coefficient ϕ_i is nearly equal to unity. Thus, the measurement of the partial pressure py_i in the vapor phase provides a measure, by means of Eq. 21.2, of the chemical potential of the liquid.

The form chosen for the liquid-phase chemical potential in Eq. 21.1 allows us to rewrite the phase–equilibrium relation in the form

$$x_i\gamma_i = py_i\phi_i \exp\left(\frac{\mu_i^* - \mu_i^\circ}{RT}\right).$$
(21.4)

The combination of $x_i\gamma_i$ is frequently referred to as the *activity* of species i. For low pressures, the relation of $\mu_i^* - \mu_i^\circ$ to the vapor pressure of pure component i is immediately evident from Eq. 21.4 since we can then take $\phi_i = 1$. Also, both x_i and y_i equal 1 for a pure component, and Eq. 21.4 becomes

$$p = p_{i,\text{sat}}^\circ = \exp\left(\frac{\mu_i^\circ - \mu_i^*}{RT}\right),$$
(21.5)

since the secondary reference state defined in Eq. 21.1 also implies that $\gamma_i = 1$ for a pure component. The vapor–pressure relationship is explored further in Example 21.1.

At higher pressures, there are several corrections that require consideration, but may obscure the development for the first exposure of readers. First, the fugacity coefficient must be determined by the methods discussed in Chapter 18. If available, vapor-phase PVT data would be used to calculate ϕ_i. Otherwise, ϕ_i must be estimated. Other factors to consider are the pressure dependences of the liquid properties μ_i° and γ_i. These are generally smaller than the pressure dependence of ϕ_i, but careful work requires their treatment.

The data in Fig. 21.1 refer to the binary system of toluene and nitroethane at 45°C, as investigated by Orye [1]. Given here are the total pressure p and the vapor-phase mole fractions y_i as a function of the liquid-phase composition, as expressed by x_1. The pressure remains near one-tenth of an atmosphere, and we are inclined to ignore the complications of high pressures. Equations 21.4 and 21.5 thus reduce to

$$py_i = p_i = x_i \gamma_i p^o_{i,\text{sat}}. \tag{21.6}$$

Equation 21.6 says that for an ideal solution, where $\gamma_i = 1$, the partial pressure in the vapor phase is strictly proportional to the mole fraction x_i in the liquid phase. This simple situation is referred to as Raoult's law and is sketched as dashed lines in Fig. 21.1. The approximate validity of Raoult's law was a motivation for expressing the chemical potential according to Eq. 21.1, with the activity coefficient γ_i describing departures from this situation. In fact, Eq. 21.1 for the liquid is very similar to Eq. 21.3 for the vapor phase. A substantial difference is that the vapor phase has a dominant pressure dependence as shown explicitly in Eq. 21.3. The condensed liquid phase has a small molar volume and is relatively insensitive to pressure changes (see Eq. 16.1 and Examples 7.3 and 9.1). Hence p does not appear explicitly in Eq. 21.1, but μ^o_i, as well as γ_i, has a weak pressure dependence.

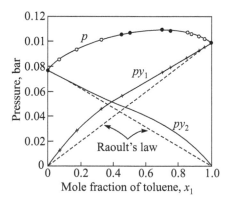

Figure 21.1 Experimental total pressure and calculated component partial pressures for liquid mixtures of toluene (component 1) and nitroethane (component 2) at 45°C.

It is evident from Fig. 21.1 that Raoult's law does not describe exactly the situation for mixtures of toluene and nitroethane. The discrepancy must here be ascribed to the composition dependence of the activity coefficient. In fact, γ_i can be calculated from the data in Fig. 21.1. The result is shown in Fig. 21.2. The activity coefficients depart appreciably from 1, but they approach 1 very closely as x_i gets close to 1. This and other features of Figs. 21.1 and 21.2 can be understood better by a more thorough study of the thermodynamics of liquid mixtures. For example, Orye does not actually measure the vapor mole fraction y_i. From measured values of the total pressure p and the application of thermodynamic principles, he calculates the vapor-phase composition.

Another interesting feature of Fig. 21.1 is the maximum in the total pressure at a liquid mole fraction of about 0.7. This indicates the presence of an *azeotrope* where the liquid and vapor compositions are identical. Although nonideal behavior is necessary to produce an azeotrope, there is nothing in Fig. 21.2 to indicate the presence of the azeotrope. If two substances have nearly equal pure-component vapor pressures, moderate departures from ideality can result in a maximum (or minimum) in the vapor pressure at an intermediate composition. The existence of an azeotrope is, therefore, related to the properties of both the vapor and liquid phases.

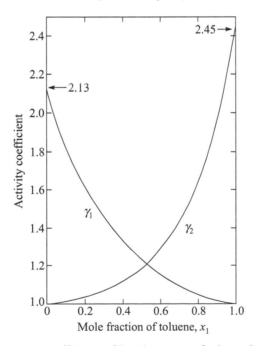

Figure 21.2 Activity coefficients of liquid mixtures of toluene (component 1) and nitroethane (component 2) at 45°C.

21.2   Gibbs Function and Ideal Solutions

For multicomponent systems, it is always wise to begin with the total Gibbs function G and thereby ensure thermodynamic consistency. Let us express G in the Margules form as

$$G = \sum_i n_i \mu_i^o + RT \sum_i n_i \ln x_i$$

$$+ \frac{1}{2} \sum_i \sum_j A_{i,j}^* n_i x_j + \frac{1}{3} \sum_i \sum_j \sum_k B_{i,j,k}^* n_i x_j x_k + \cdots, \qquad (21.7)$$

where the parameters $A_{i,j}^*$ and $B_{i,j,k}^*$ depend on temperature and pressure but are independent of composition. The first term can be regarded as the contribution of the secondary reference states—the pure components. The remaining terms must be zero for a pure component. For this purpose, we require that

$$A_{i,i}^* = 0, \qquad (21.8)$$

$$B_{i,i,i}^* = 0, \text{ etc.} \qquad (21.9)$$

(Additional trivial restrictions can also be placed on these coefficients; see Eqs. 21.20 to 21.22.) The second term on the right in Eq. 21.7 is the ideal Gibbs function of mixing and can be compared with Eq. 17.21 for an ideal-gas mixture. The remaining terms in Eq. 21.7 describe departures from ideal behavior and will be the subject of subsequent discussion.

Frequently, the *excess Gibbs function* G^E is defined as

$$G^E = G - \sum_i n_i \mu_i^o - RT \sum_i n_i \ln x_i \qquad (21.10)$$

and thereby represents all the nonideal terms in Eq. 21.7. The excess Gibbs function can then be expressed in a manner different from the Margules form used in Eq. 21.7. One must always remember clearly what was subtracted from G in order to get G^E according to Eq. 21.10, since similar equations apply to G for gas mixtures (Eqs. 18.2 and 18.3) and for dilute solutions (Eqs. 20.19 and 20.31).

For an *ideal solution*, the terms $A_{i,j}^*$ and $B_{i,j,k}^*$ in Eq. 21.7 are zero, and the total Gibbs function is simply

$$G = \sum_i n_i \mu_i^o + RT \sum_i n_i \ln x_i . \tag{21.11}$$

Differentiation of G according to Eq. 14.14 yields the chemical potential[2]

$$\mu_i = \mu_i^o + RT \ln x_i + RT \sum_j n_j \left(\frac{\delta_{i,j}}{n_j} - \frac{1}{n} \right), \tag{21.12}$$

where $\delta_{i,j}$ is the Kronecker delta having the properties

$$\begin{aligned} \delta_{i,j} &= 1 \text{ if } i = j, \\ &= 0 \text{ if } i \neq j. \end{aligned} \tag{21.13}$$

The last terms on the right cancel, and there remain

$$\mu_i = \mu_i^o + RT \ln x_i . \tag{21.14}$$

Comparison with Eq. 21.1 shows that the activity coefficient γ_i is indeed unity for an ideal solution. It is thus equivalent to say that an ideal solution is defined by $\gamma_i = 1$ for all components or by the expression of the Gibbs function according to Eq. 21.11.

Differentiation of Eq. 21.11 with respect to pressure gives

$$\left(\frac{\partial G}{\partial p} \right)_{T,n_j} = V = \sum_i n_i \left(\frac{\partial \mu_i^o}{\partial p} \right)_T = \sum_i n_i \tilde{V}_i^o , \tag{21.15}$$

where $\tilde{V}_i^o$ is the molar volume of the pure liquid component. Comparison with Eqs. 14.11 and 14.17 shows that the volume change on mixing is zero for an ideal solution. Furthermore, since $\tilde{V}_i^o$ is independent of composition, Eq. 21.15 shows that the partial molar volumes are independent of composition,

$$\overline{V}_i = \tilde{V}_i^o, \tag{21.16}$$

for an ideal solution.

Differentiation of Eq. 21.11 with respect to temperature gives

$$\left(\frac{\partial G}{\partial T} \right)_{p,n_j} = -S = \sum_i n_i \left(\frac{\partial \mu_i^o}{\partial T} \right)_p + R \sum_i n_i \ln x_i \tag{21.17}$$

or

$$S = \sum_i n_i \tilde{S}_i^o (T,p) - R \sum_i n_i \ln x_i . \tag{21.18}$$

---

[2]Readers may have an aversion to differentiation of complicated, multicomponent summations. Faith is an acceptable alternative.

The last term is recognized as the ideal entropy of mixing. From Eqs. 21.11, 21.18, 7.16, and 14.18, one can show that the enthalpy change on mixing is zero and that the partial molar enthalpies are independent of composition for an ideal solution.

Vapor pressures and thermodynamic diagrams for pure components were treated in Chapter 11, and several sources of data are cited. The chemical potential μ_i^o of a pure liquid is related to its vapor pressure through the condition of phase equilibrium (compare Eqs. 21.4 and 21.5):

$$\mu_i^o(T,p) = \mu_i^*(T) + RT \ln(p_{sat}\phi_{i,sat}^o) + (p - p_{sat})\tilde{V}_i^o , \quad (21.19)$$

where it has been assumed that the liquid molar volume $\tilde{V}_i^o$ is independent of pressure. Here, μ_i^* can be obtained by integrating the ideal-gas heat capacity as discussed in Example 17.4 and in Section 19.3, and the fugacity coefficient at the vapor pressure can be obtained by the methods of Chapter 18. Alternatively, one can begin with the Gibbs function and enthalpy of the pure liquid at some reference temperature (see Table 19.2) and integrate the liquid-phase heat capacity. (This is also done in Example 21.1.)

21.3 Margules Equations

Departures from ideal behavior can be described by adding correction terms to the ideal Gibbs function in Eq. 21.11. According to the method of Margules, the correction can take the form of a power-series expansion in the mole fractions as in Eq. 21.7. Inspection of that equation shows that without loss of generality, we can set

$$A_{i,j}^* = A_{j,i}^* , \quad (21.20)$$

since $n_i x_j$ is identical to $n_j x_i$. Thus, there is one independent parameter $A_{i,j}^*$ corresponding to each pair of components in the mixture, a total of $(C^2 - C)/2$ such parameters. Similar reasoning shows that we can set

$$B_{i,j,k}^* = B_{k,i,j}^* = B_{j,k,i}^* = B_{i,k,j}^* = B_{j,i,k}^* = B_{k,j,i}^* . \quad (21.21)$$

This would seem to reduce the number of B^* values by a factor of 6, except when two of the subscripts are the same and this relation equates only 3 B^* values. However, we must also bear in mind that the mole fractions are not independent in a mixture; they must sum to unity. This interdependence allows us also to write

$$B^*_{i,j,j} = -B^*_{i,i,j}. \qquad (21.22)$$

The restrictions of this paragraph are in addition to those expressed in Eqs. 21.8 and 21.9. There are $(C^2 - C)/2$ independent parameters of the form $B^*_{i,i,j}$ corresponding to the number of binary pairs in the mixture. In addition, for each of $C(C - 1)(C - 2)/6$ ternary systems, which can be formed from the components in the mixture, there is a parameter of the form $B^*_{i,j,k}$ where the subscripts are distinct. This leads to a total of $(C^3 - C)/6$ independent B^* values for a mixture.

Thus, for a binary system, there are 1 A^* value and 1 B^* value, and the resulting equations are referred to as *three-suffix* Margules equations when terms beyond the B^* terms in Eq. 21.7 are omitted. For a ternary system, there are 3 A^* values and 4 B^* values. All the A^* values and 3 of the B^* values can be obtained by studies of the binary systems that can be formed from the components of the mixture. Thus, only one parameter, $B^*_{1,2,3}$ needs to be determined from mixtures containing all three components simultaneously. For mixtures of more than three components, all the required A^* and B^* values can be obtained by measurements on the simpler binary and ternary systems that can be formed from the components.

We shall assume that terms beyond those shown explicitly in Eq. 21.7 are seldom used to fit thermodynamic properties in multicomponent systems. This is because of the complexity of the undertaking and the scarcity of good data, not because the included terms are always adequate. Binary systems are an exception. Here very good data can sometimes be found, and the use of terms through C^* and D^* gives a four-parameter fit of the data (so called *five-suffix* Margules equations).

Values of Margules coefficients that have been used for some binary systems are presented in Table 21.1. The additional $B^*_{1,2,3}$ value necessary to treat properly a ternary system is given for a few cases in Table 21.2. Here a zero value means that the ternary system has been studied experimentally and the binary coefficients in Table 21.1 have been adequate to reproduce the data.

Table 21.1 Binary Margules coefficients.

Component 1	Component 2	Temperature	$A^*_{1,2}/RT$	$B^*_{1,1,2}/RT$
argon	oxygen	83.8 K	0.213	0
		89.6 K	0.189	0
benzene	cyclohexane	30	0.503	0
		40	0.445	0
		50	0.414	0
n-hexane	toluene	1 atm	0.352	0
benzene	isooctane	~1 atm	0.433	0.104
acetone	chloroform	50	−0.760	0.07
acetone	methanol	50	0.610	−0.091
chloroform	methanol	50	1.2575	0.5425
chloroform	ethanol	55	1.005	0.415
benzene	cyclopentane	25	0.45598	−0.01815
		35	0.42463	−0.01627
		45	0.40085	−0.02186
acetone	methyl acetate	50	0.132	−0.017
acetone	carbon tetrachloride	50	0.830	0.115
methyl acetate	methanol	50	1.045	−0.025
carbon tetrachloride	methanol	50	2.14	0.38

Note: Temperatures are in degrees Celsius unless otherwise indicated.

Table 21.2 Ternary Margules coefficients.

Components	Temperature °C	$B^*_{1,2,3}/RT$
acetone/methyl acetate/methanol	50	0
acetone/chloroform/methanol	50	0.184
acetone/carbon tetrachloride/methanol	50	−0.575

For a ternary mixture, the excess Gibbs function G^E takes the form

$$G^E = A_{12}^* n_1 x_2 + A_{13}^* n_1 x_3 + A_{23}^* n_2 x_3 + B_{112}^* n_1 x_2 (x_1 - x_2)$$

$$+ B_{113}^* n_1 x_3 (x_1 - x_3) + B_{223}^* n_2 x_3 (x_2 - x_3) + 2B_{123}^* n_1 x_2 x_3, \quad (21.23)$$

and the activity coefficient of a component follows from Eq. 21.7 by differentiation according to Eq. 14.14. For example, for component 1,

$$RT \ln \gamma_1 = A_{12}^* x_2 (1 - x_1) + A_{13}^* x_3 (1 - x_1) - A_{23}^* x_2 x_3$$

$$+ B_{112}^* x_2 (2x_1 - x_2 - 2x_1^2 + 2x_1 x_2)$$

$$+ B_{113}^* x_3 (2x_1 - x_3 - 2x_1^2 + 2x_1 x_3)$$

$$- 2B_{223}^* x_2 x_3 (x_2 - x_3) + 2B_{123}^* x_2 x_3 (1 - 2x_1). \quad (21.24)$$

Expressions for the activity coefficients of the other components can be obtained from this equation by interchanging the subscripts. For this purpose, and in other applications, recall that $B_{233}^* = -B_{223}^*$, etc. For a binary mixture of components 1 and 2, Eq. 21.24 simplifies to

$$RT \ln \gamma_1 = A_{12}^* x_2^2 + B_{112}^* x_2^2 (4x_1 - 1). \quad (21.25)$$

Figure 21.3 shows the activity coefficients of the components in binary mixtures of benzene and cyclohexane at 40°C. Here $B_{112}^* = 0$, and the system is an example of a *symmetric mixture*. The curves have the same shape, and the activity coefficient of component 1 in a large excess of component 2 ($x_1 \to 0$) is equal to the activity coefficient of component 2 in a large excess of component 1 ($x_1 \to 1$). Two other characteristics of the curves in Fig. 21.3 are more general. The activity coefficient of component i approaches unity in pure component i because of the selection of the secondary reference state in Eq. 21.1. Furthermore, the slope is zero here, as can be seen in the presence of the factor x_2^2 in Eq. 21.25. The activity coefficient is very close to unity for a component present in large excess, and this was the reason for using the osmotic coefficient in Chapter 20 to represent the nonideal behavior of the chemical potential of the solvent in a dilute solution. This is also why the partial-pressure curves in Fig. 21.1 become tangent to the lines for Raoult's law as the mole fraction of the other component approaches zero.

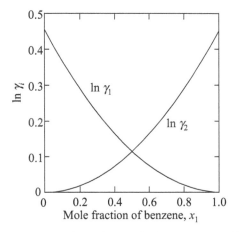

Figure 21.3 Composition dependence of the activity coefficients in the symmetric mixtures of benzene and cyclohexane at 40°C.

Figure 21.4 shows the values of G^E for this symmetric system. Of more direct physical significance, perhaps, is the Gibbs function change on mixing. This is still dominated by the ideal terms (the second term on the right in Eq. 21.7), although addition of G^E has lowered somewhat the curvature near $x_1 = 0.5$. The shape of the curve for the Gibbs function change on mixing governs the tendency of the system to split into two liquid phases, as discussed in Example 21.3.

In the simplest of the lattice or chemical theories of liquid mixtures, the $A^*_{i,j}$ parameter can be interpreted as the potential energy (at the average separation distances prevailing in the liquid) of an i,j pair relative to those of i,i and j,j pairs:

$$2A^*_{i,j} = z(2\varepsilon_{i,j} - \varepsilon_{i,i} - \varepsilon_{j,j}),\tag{21.26}$$

where the energies are expressed here on a molar basis and z is the number of nearest neighbors for a given molecule.

Physical interactions between the molecules usually lead to positive values of $A^*_{i,j}$ and to positive deviations from Raoult's law (as illustrated in Fig. 21.1). This means that like molecules have a stronger attraction than unlike molecules. Inspection shows that most of the $A^*_{i,j}$ values in Table 21.1 are positive. For a given pair, the values decrease with increasing temperature (and the tabulated values of $A^*_{i,j}/RT$ decrease even faster). This decrease could be attributed to the weaker interactions at the greater separation distances prevailing at higher temperatures.

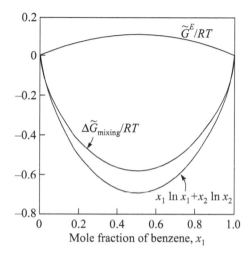

Figure 21.4 Excess molar Gibbs function for symmetric mixture of benzene and cyclohexane at 40°C. Also shown on the same scale are the ΔG of mixing (middle curve) for this system and the ideal ΔG of mixing (bottom curve).

For large values of $A^*_{i,j}$, the relatively stronger attraction of like molecules can lead to the formation of two liquid phases. This means that, for a given overall composition, two phases have a lower total Gibbs function than a single phase at the same temperature and pressure. (Compare with the condition of equilibrium in Fig. 11.4d and the discussion of phase equilibrium between a gas and a liquid in Figs. 11.5 and 11.6 for a van der Waals fluid.) The condition of incomplete miscibility begins at $A^*_{i,j} = 2RT$ when the B^* value is zero,[3] and the usual decrease in $A^*_{i,j}/RT$ with increasing temperature can explain the usual increase with temperature of the mutual solubility of two immiscible liquids.

Specific chemical attractions produce negative deviations from Raoult's law. Acetone and chloroform show a negative value of $A^*_{1,2}$ in Table 21.1. Here the stability of the mixture is enhanced by hydrogen bonding between the unlike molecules. Figure 21.5 shows

---

[3]With $A^*_{i,j} > 2RT$ for carbon tetrachloride and methanol at 50°C (see Table 21.1), this system is incompletely miscible even with the prevailing value B^*_{112} (see also Example 21.3).

the activity coefficients, which are less than unity due to the chemical interactions.

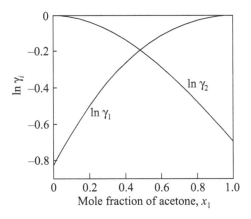

Figure 21.5 Composition dependence of the activity coefficients in mixtures of acetone and chloroform, a system with negative deviations from Raoult's law. The temperature is 50°C.

The Margules formulation has the advantage that its application to multicomponent systems is straightforward. Data obtained on binary systems can be used to simplify the treatment of a ternary mixture, and predictions can be made if one is willing to neglect the B^*_{123} term. This method has the disadvantage that it defines a multitude of parameters with unspecified dependences on temperature and pressure. It is not clear that the straightforward expansion in mole fractions has a physical basis which would suggest that actual thermodynamic data can be well fitted by a Margules expression with a minimum of adjustable parameters.

For binary systems, there are several variants for expansion of the excess Gibbs function as a power series in the mole fractions. One form reads

$$\frac{\widetilde{G}^E/RT}{x_1 x_2} = a + b x_1 + c x_1^2 + \cdots. \tag{21.27}$$

The Redlich–Kister expansion can be written as

$$\frac{\widetilde{G}^E/RT}{x_1 x_2} = B + C(x_1 - x_2) + D(x_1 - x_2)^2 + \cdots. \tag{21.28}$$

A special truncated form for binaries is

$$\frac{\tilde{G}^E/RT}{x_1 x_2} = A_{21}x_1 + A_{12}x_2 .$$ (21.29)

On the other hand, the standard form (from Eq. 21.23) can be written as

$$\frac{\tilde{G}^E/RT}{x_1 x_2} = \frac{A_{12}^*}{RT} + \frac{B_{112}^*}{RT}(x_1 - x_2).$$ (21.30)

Since these four expressions should be equivalent, there must be relationships among the coefficients so defined. The Redlich–Kister form is quite similar to the standard form, and the equivalence leads to

$$B = A_{12}^*/RT \quad \text{and} \quad C = B_{112}^*/RT .$$ (21.31)

If the standard form is re-expressed as

$$\frac{\tilde{G}^E/RT}{x_1 x_2} = \frac{A_{12}^* + B_{112}^*}{RT}x_1 + \frac{A_{12}^* - B_{112}^*}{RT}x_2 ,$$ (21.32)

it follows that

$$A_{21} = \frac{A_{12}^* + B_{112}^*}{RT} \quad \text{and} \quad A_{12} = \frac{A_{12}^* - B_{112}^*}{RT}.$$ (21.33)

The inverse relationship is

$$\frac{A_{12}^*}{RT} = \frac{A_{21} + A_{12}}{2} \quad \text{and} \quad \frac{B_{112}^*}{RT} = \frac{A_{21} - A_{12}}{2}.$$ (21.34)

Finally, Eq. 21.30 can be rewritten as

$$\frac{\tilde{G}^E/RT}{x_1 x_2} = \frac{A_{12}^* - B_{112}^*}{RT} + \frac{2B_{112}^*}{RT}x_1 ,$$ (21.35)

and comparison with Eq. 21.27 leads to the equivalence

$$a = A_{12} = \frac{A_{12}^* - B_{112}^*}{RT} \quad \text{and} \quad b = \frac{2B_{112}^*}{RT}.$$ (21.36)

The solution for A_{12}^* gives

$$\frac{A_{12}^*}{RT} = a + \frac{b}{2}.$$ (21.37)

Holmes and van Winkle [2] and Mertl [3] provide values of Margules parameters for many binary systems. Their tables are much more extensive than Table 21.1.

21.4   Van Laar Equation and Regular Solutions

The Margules form expresses the excess Gibbs function as a power-series expansion in the mole fractions x_i. Wohl's expansion is a generalization where powers of z_i are used, where

$$z_i = x_i q_i / q \tag{21.38}$$

and

$$q = \sum_i x_i q_i . \tag{21.39}$$

The parameters q_i can depend on temperature and pressure, or they can be constants. A frequent choice is to let q_i equal the molar volume of pure component i

$$q_i = \tilde{V}_i^o (T, p) \tag{21.40}$$

or the same quantity but evaluated at the vapor pressure

$$q_i = \tilde{V}_i^o (T, p_{i,\text{sat}}^o) \tag{21.41}$$

or at some reference pressure. In any event, Wohl's expansion takes the form

$$G = \sum_i n_i \mu_i^o + RT \sum_i n_i \ln x_i + \frac{1}{2} \sum_i \sum_j A_{ij} q_i n_i z_j$$
$$+ \frac{1}{3} \sum_i \sum_j \sum_k B_{ijk} q_i n_i z_j z_k + \cdots . \tag{21.42}$$

The intensive version would be

$$\frac{\tilde{G}^E}{RTq} = \frac{1}{2} \sum_i \sum_j \frac{A_{ij}}{RT} z_i z_j + \frac{1}{3} \sum_i \sum_j \sum_k \frac{B_{ijk}}{RT} z_i z_j z_k + \cdots . \tag{21.43}$$

Wohl's expansion actually includes the Margules form as a special case, when $q_i = 1$ for all i. Then $A_{ij} = A_{ij}^*$, etc. If enough terms are carried, either form could give an adequate fit of the Gibbs function. Preference for one form or another may derive from how few parameters will suffice with truncated versions, what form of parameters is readily available, and how well one can predict the parameters from limited data. The last criterion deals with the perennial problem of treating a ternary system or multicomponent system based on data for binary systems or even from pure-component data alone.

Only ratios of q_i values are important, since any other effect can be incorporated into the A's, B's, etc. Also, since the A's, B's, etc. provide enough freedom to fit any G function, the *a priori* choice of q_i is justified. That is why the choice $q_i = \tilde{V}_i^o(T,p)$ may be made.

Not many people deal with the general form of Wohl's expansion, but several popular forms are special cases of it, and the author would be sad if he had not made it evident how binary data fitted by, say, van Laar's equation might be extended to a multicomponent formulation with or without needing to refit parameters to the binary data.

The fact that G^E is zero for a pure component leads to conditions analogous to Eqs. 21.8 and 21.9 on the coefficients. Symmetry considerations and the fact that the z_i values sum to unity lead to restrictions analogous to Eqs. 21.20–21.22. Familiarity with the Margules form should be useful here.

For binary mixtures, the van Laar equations have been popular in treating the data. These read

$$\ln \gamma_1 = \frac{A'_{12}}{\left(1 + \dfrac{A'_{12}x_1}{A'_{21}x_2}\right)^2} \quad \text{and} \quad \ln \gamma_2 = \frac{A'_{21}}{\left(1 + \dfrac{A'_{21}x_2}{A'_{12}x_1}\right)^2}. \quad (21.44)$$

You might ask yourself whether these forms are thermodynamically consistent and how you should go about answering such a question. These forms for the binary activity coefficients can be derived from Wohl's expansion if we identify A'_{12} and A'_{21} as

$$A'_{12} = A_{12}q_1/RT \quad \text{and} \quad A'_{21} = A_{12}q_2/RT, \quad (21.45)$$

and terms beyond A_{ij} are neglected in Wohl's expansion. (See also Problems 21.5 and 21.8.) Activity coefficients fitted to the van Laar equations do not differ dramatically from what we are used to; Fig. 21.6 shows the activity coefficients for mixtures of water and ethanol.

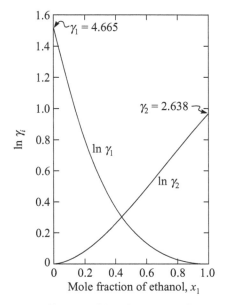

Figure 21.6 Activity coefficients of liquid mixtures of ethanol (component 1) and water (component 2) at 25°C, based on the van Laar equation.

We have collected some values of van Laar parameters in Table 21.3. (See also Ref. [2].) The best fits of A'_{12} and A'_{21} to binary data cannot necessarily be used in multicomponent systems unless one looks ahead and ensures that a single value is used for q_i for a particular species i when analyzing all binary pairs involving this species. (See also Problem 21.7.) It might be better to refit all the binary data with the two-parameter form that results when terms in the B's are retained (see Problem 21.11). Then information from binary systems could be extended to treat multicomponent systems in a manner completely analogous to the treatment of the three-suffix Margules equation.

Table 21.3 Coefficients for the binary van Laar equation.

Component 1	Component 2	A'_{12}	A'_{21}	Condition
methanol	ethyl acetate	1.332	0.985	40°C
		1.031	1.243	50
		1.098	1.091	60
acetone	chloroform	−0.701	−0.624	1 atm

Component 1	Component 2	A'_{12}	A'_{21}	Condition
benzene	chloroform	−0.198	−0.128	1 atm
ethyl acetate	chloroform	−1.031	−0.660	1 atm
water	acetone	1.406	2.296	1 atm
water	acetonitrile	1.890	2.459	1 atm
water	ethanol	0.945	1.679	1 atm
		0.937	1.676	75° C
		0.923	1.718	87.8°C
		0.904	1.751	100°C
water	methanol	0.562	0.889	1 atm
water	1-propanol	1.160	2.633	1 atm
		1.222	2.836	40°C
		1.203	2.735	60°C
ethyl acetate	ethanol	0.956	1.113	40°C
		0.885	0.942	60°C
ethanol	methanol	0.020	0.058	1 atm
ethyl acetate	1-propanol	0.793	0.901	40°C
		0.636	0.712	60°C
ethyl acetate	2-propanol	0.863	1.060	40°C
		0.700	0.751	60°C
acetone	methanol	0.636	0.663	55°C
acetone	2-propanol	0.727	0.574	55°C

It is always nice if one can make *a priori* predictions of parameters to use in equations, and this is what van Laar did. To be done properly, the rigorous methods of statistical mechanics should be used. However, this is a difficult undertaking and may be impossible to do without some approximations. The objective would be to relate thermodynamic properties to the properties of the molecules themselves. In the absence of a rigorous answer, there is a lot of room for creativity in developing approximations with more or less physical appeal and more or less accuracy. As described by Prausnitz (Ref. [4], pp. 264–269), van Laar made several simplifying assumptions:

1. There is no volume change on mixing.

2. The entropy of mixing is given by the ideal formula. (See Problem 21.1.)
3. The pure components obey the van der Waals equation of state.
4. The mixture obeys the van der Waals equation of state with parameters a and b related to the pure component values by the mixing rules

$$\sqrt{a} = x_1\sqrt{a_1} + x_2\sqrt{a_2} \quad \text{and} \quad b = x_1 b_1 + x_2 b_2. \qquad (21.46)$$

It follows from the first two assumptions that the excess Gibbs function is equal to the internal energy of mixing, that is, the non-ideality can be associated solely with an energy effect. Van Laar then evaluated the internal energy of mixing by lowering the pressure on the pure components, mixing at low pressure where the Gibbs mixing rule applies, and then repressurizing the system to the desired pressure. He also took the molar volume of the liquid to be equal to the van der Waals parameter b.

In this way van Laar arrived at the activity coefficients for the mixture, as well as the excess Gibbs function, in a way that can be expressed by the parameter relations

$$q_1 = b_1, q_2 = b_2, \text{and } A_{12} = \left(\frac{\sqrt{a_1}}{b_1} - \frac{\sqrt{a_2}}{b_2} \right)^2, \qquad (21.47)$$

and $B_{ijk} = 0$ in the Wohl's expansion. Other, substantially different, parameter relations could have been obtained by assuming different mixing rules for a and b. The values of a_1, b_1 a_2, and b_2 could be obtained from the critical properties, as discussed in Chapter 3, or values could be selected by fitting the pure-component vapor–pressure curve over a range closer to the operating conditions.

Working independently of each other, Scatchard and Hildebrand worked to relax the assumption that the internal energy of mixing should be determined from the van der Waals equation of state while substantially retaining the assumptions that the volume change on mixing be zero and the excess entropy be zero. The mixtures thus treated are termed regular solutions. Their result can then be expressed by means of the parameter relations

$$q_i = \tilde{V}_i^o \text{ and } A_{ij} = (\delta_i - \delta_j)^2 \qquad (21.48)$$

where δ_i is called the *solubility parameter* defined as the square root of the difference between the internal energy in the ideal gas and in the pure liquid, divided by the liquid molar volume:

$$\delta_i = \left(\frac{\tilde{U}_i^* - \tilde{U}_i^0}{\tilde{V}_i^0} \right)^{1/2}. \tag{21.49}$$

Some values of $\tilde{V}_i^0$ and δ_i are given in Table 21.4.

Table 21.4 Solubility parameters, pure-component vapor pressure, and liquid molar volumes for selected materials at 25°C.

Substance	δ_i (cal/cm³)$^{1/2}$	$\tilde{V}_i^0$ (cm³/mol)	$p_{i,sat}^0$ (bar)
methanol	—	40.6	0.168
carbon disulfide	10.0	61	0.333
acetone	—	74	0.305
chloroform	9.3	81	0.265
cyclohexane	8.20	109	0.1303
benzene	9.15	89	0.1268
carbon tetrachloride	8.6	97	0.153
ethyl ether	7.4	105	0.716
stannic chloride	8.7	118	0.0320
n-heptane	7.45	147	0.0600
mercury	—	14.8	2.45 × 10⁻⁶
bromine	11.5	51	—
propane	6.0	89	—
n-butane	6.7	101	—
isobutene	6.25	105	—
n-pentane	7.05	116	—
toluene	8.90	107	—
bromoform	10.5	88	—
nitromethane	12.6	54	—

Source: Joel H. Hildebrand and Robert L. Scitt, *The Solubility of Nonelectrolytes* (New York: Dover Publications, Inc., 1964), pp. 76, 435–439.

Evaluation of A_{ij} values according to either the original method of van Laar or in terms of solubility parameters leads always to positive values and to positive deviations from Raoult's law. (The van Laar equation itself can describe negative deviations from Raoult's law; see the entries in Table 21.3 for the system of acetone and chloroform.) Only pure component properties are needed to use regular-solution theory, and larger departures from ideality are predicted for mixtures of substances that are dissimilar, as measured by the solubility parameters. The predictions are considered to be useful for nonpolar molecules, although one generally prefers to use parameters actually fitted to binary data if these are available. This prediction method is considered quite unreliable if one of the molecules is *polar*.

Although values of δ_i and $\tilde{V}_i^o$ depend on temperature, differences of δ_i and ratios of $\tilde{V}_i^o$ are not very sensitive to temperature, and the values used should preferably all be evaluated at the same temperature, although perhaps one different from the operating temperature. The constant values of A_{ij} are in harmony with the assumption of a zero excess entropy (see Problem 21.1).

21.5 Wilson Equation

Another expression for the excess Gibbs function, due to Wilson [5], takes the form

$$\tilde{G}^E / RT = -\sum_i x_i \ln\left(\sum_j x_j \Lambda_{ij}\right). \qquad (21.50)$$

For each pair of species in the mixture, there are two parameters, Λ_{ij} and Λ_{ji}, which are generally not equal to each other. However, $\Lambda_{i,i}$ = 1 in order that $\tilde{G}^E$ might vanish for pure components. One of the principal advantages of the Wilson equation is already apparent: the complete thermodynamics of multicomponent solutions can be predicted on the basis of data on binary systems alone. This is a great benefit in the absence of detailed data, and it should be straightforward to handle in a computer. A disadvantage also lurks here; there is no way to accommodate data for ternary mixtures even if such data are available. One must accept the predictions from binary data even if they are known to be wrong.

In keeping with the general rules for parameters occurring in expressions for the excess Gibbs function, Λ_{ij} can be a general function of temperature and pressure. The recommended form is

$$\Lambda_{ij} = \frac{\tilde{V}_j^o}{\tilde{V}_i^o} \exp\left(-\frac{a_{ij}}{RT}\right), \qquad (21.51)$$

where a_{ij} is a parameter independent of temperature and pressure (and, again, a_{ij} is generally not equal to a_{ji}). The ratio $\tilde{V}_j^o / \tilde{V}_i^o$ of the molar volumes of the pure components can also be taken to be independent of temperature and pressure to a good and useful approximation. With these forms for Λ_{ij}, the Wilson equation provides a powerful approach to correlating and predicting thermodynamic properties and expressing them in a form suitable for machine computation. Parameter values for selected binary pairs are given in Table 21.5. Holmes and van Winkle [2] provide another tabulation. Some molar volumes can be found in Table 21.4.

Table 21.5 Parameters for use in the Wilson equation, together with the temperature or pressure for which they were determined from vapor–liquid equilibria.

Component 1	Component 2	Λ_{12}	Λ_{21}	Condition
toluene	acetonitrile	0.43768	0.53776	45°C
toluene	2,3-butanedione	0.66708	0.61216	45°C
toluene	acetone	0.68363	0.70513	45°C
toluene	nitroethane	0.80959	0.50507	45°C
methyl-cyclohexane	acetone	0.31554	0.30235	45°C
		a_{12} (kJ/mol)	a_{21} (kJ/mol)	
ethanol	benzene	6.853	0.478	45°C
		6.402	0.481	1 atm
n-butanol	benzene	5.115	0.747	45°C
n-propanol	benzene	5.669	0.785	45°C
acetone	benzene	2.059	−0.663	45°C
nitromethane	benzene	3.124	0.599	45°C
		2.986	0.682	25°C

(Continued)

Table 21.5 *(Continued)*

Component 1	Component 2	a_{12} (kJ/mol)	a_{21} (kJ/mol)	Condition
acetonitrile	benzene	3.512	−0.381	45°C
nitroethane	benzene	2.020	−0.040	25°C
1-nitropropane	benzene	2.637	−0.914	25°C
2-nitropropane	benzene	2.746	−0.988	25°C
methyl-cyclopentane	benzene	0.055	1.041	1 atm
n-hexane	benzene	0.647	0.890	1 atm
n-heptane	benzene	1.456	0.517	60°C
benzene	iso-octane	0.844	1.161	1 atm
cyclohexane	ethanol	1.865	9.661	5°C
		1.773	9.101	20°C
		1.702	8.694	35°C
		1.592	8.578	50°C
		1.397	8.377	65°C
ethanol	methyl-cyclopentane	9.297	1.041	1 atm
n-hexame	methyl-cyclopentane	−0.368	0.431	1 atm
ethanol	n-hexane	8.984	1.369	1 atm

Note: Λ_{ij} values are given for the first five systems; a_{ij} values for the remainder.

Corresponding to Eq. 21.50, the activity coefficient is found (by differentiation according to Eq. 14.14) to be

$$\ln \gamma_k = -\ln\left(\sum_j x_j \Lambda_{kj}\right) + 1 - \sum_i \frac{x_i \Lambda_{ik}}{\sum_j x_j \Lambda_{ij}}. \qquad (21.52)$$

For binary mixtures, these equations simplify to

$$\tilde{G}^E / RT = -x_1 \ln(x_1 + \Lambda_{12} x_2) - x_2 \ln(x_2 + \Lambda_{21} x_1) \qquad (21.53)$$

and

$$\ln \gamma_1 = -\ln(x_1 + \Lambda_{12} x_2) + x_2 \left(\frac{\Lambda_{12}}{x_1 + \Lambda_{12} x_2} - \frac{\Lambda_{21}}{\Lambda_{21} x_1 + x_2}\right). \qquad (21.54)$$

21.6   Phase Equilibria

The equilibrium between a liquid and its vapor has already been treated for single components. The vapor–pressure curve is given in Figs. 11.3 and 11.7, and the temperature dependence of the vapor pressure is described by Eq. 11.12. The condition of equilibrium is expressed in Eq. 11.9. This result applies also for multicomponent systems; for each component present in both phases, the chemical potential is equal.

The development of thermodynamic properties of solutions requires reversible mixing experiments. The stipulation of reversibility puts severe restrictions on the experimenter; it is easy to mix components irreversibly. But the definition of the entropy specifies reversible heat effects. The three main tools are

1. The theory of mixing in the ideal-gas limit, where the Gibbs mixing rule specifies that the volume change on mixing is zero at constant pressure and temperature. (See Figs. 14.5, 17.1, and 17.2) Because the entropy at higher pressures can be obtained by PVT properties and pressures (see Eqs. 7.7 and 7.28), this is sufficient in principle.

2. Carrying out chemical reactions in an electrochemical cell. For example, the Gibbs function of water (relative to the elements) was carried out by Gilbert Newton Lewis by a series of experiments involving several oxides. Thus, the standard-state potential for the formation of H_2O from H_2 and O_2 was deduced to be 1.229 V at 25°C [6]. Some electrochemical cells suitable for reversible mixing experiments are shown in Figs. 17.3 and 19.5.

3. Phase equilibria permit the composition of a phase to be changed reversibly in a straightforward manner, which is evident even for pure substances.

Thus, phase equilibria provide first a means for developing the chemical potentials of components in multicomponent solutions and are equally important as the Gibbs mixing rule and electrochemical reactions. And, we assert that you are already familiar with phase equilibria. See, for example, the fitting of vapor pressure data for solutions of toluene and nitroethane in Fig. 21.1. Other examples include Figs. 11.3, 11.4, and 18.2.

We do not repeat everything that can be said about phase equilibria here. It is more interesting to have already developed the material in the context of the preceding chapters. However, we should point out the huge practical significance of phase equilibria in the chemical industry, where separation of chemical components by distillation, liquid-liquid extraction, ion exchange, pressure-swing adsorption, etc. are studied in their own right in courses on separation.

Example 21.1 Reconcile thermodynamic data for water in Tables 19.1 and 19.2 with accepted experimental values for the vapor pressure of 23.753 mm Hg or 0.031254 atm at 25°C and 42.180 mm Hg or 0.0555 atm at 35°C.

Solution: Equation 21.19 relates the vapor pressure p_{sat} to the secondary reference states for the ideal gas and the pure liquid. At the low pressures involved, we can set $\phi_{i,sat}$ equal to 1. We also ignore the last term in Eq. 21.19 on the basis of the smallness of the liquid molar volume and the fact that the pressure difference is less than 1 atm. At 25°C, Tables 19.1 and 19.2 give

$$\mu_i^*(T) = -54.634 \text{ kcal/mol}$$

and

$$\mu_i^o(T, p = 1 \text{ atm}) = -56.687 \text{ kcal/mol} .$$

The vapor pressure from Eq. 21.19 is now

$$p_{sat} = \exp\left(\frac{\mu_i^o - \mu_i^*}{RT}\right) = \exp\left(\frac{-56.687 + 54.634}{1.987 \times 298.15} 10^3\right) = 0.031260 \text{ atm},$$

in good agreement with the accepted value. (Recall that the atmosphere unit is concealed in μ_i^*.)

To compare with the literature value at 35°C, we should first calculate μ_i^* and μ_i^o at that temperature. Tables 19.1 and 19.2 give us enthalpy and heat capacity values at $T_o = 25°C$:

$$\widetilde{H}_i^*(T_o) = -57.796 \text{ kcal/mol}, \quad \widetilde{H}_i^o(T_o, p = 1 \text{ atm}) = -68.315 \text{ kcal/mol},$$

$$\widetilde{C}_{pi}^*(T_o) = 8.025 \text{ cal/mol-K}, \quad \widetilde{C}_{pi}^o(T_o, p = 1 \text{ atm}) = 17.995 \text{ cal/mol-K}.$$

If we take the heat capacity to be constant and equal to a_i, we can now use the formula in Example 17.4 to estimate the value of μ_i^* at $T = 35°C$:

$$\mu_i^*(T) = \frac{T}{T_o}\mu_i^*(T_o) + \left(1 - \frac{T}{T_o}\right)\tilde{H}_i^*(T_o) - a_i\left(T\ln\frac{T}{T_o} + T_o - T\right)$$

$$= \frac{308.15}{298.15}(-54.634) + \left(1 - \frac{308.15}{298.15}\right)(-57.796)$$

$$- \frac{8.025}{10^3}\left(308.15\ln\frac{308.15}{298.15} - 10\right)$$

$$= -54.528 - 0.001331 = -54.529 \text{ kcal/mol}.$$

The term involving the heat capacity makes a very small contribution over this small temperature range.

A similar equation can be applied to the chemical potential μ_i^o of the pure liquid:

$$\mu_i^o(T, p=1\text{ atm}) = \frac{T}{T_o}\mu_i^o(T_o, p=1\text{ atm}) + \left(1 - \frac{T}{T_o}\right)\tilde{H}_i^o(T_o, p=1\text{ atm})$$

$$-\tilde{C}_{pi}^o(T, p=1\text{ atm})(T\ln\frac{T}{T_o} + T_o - T)$$

$$= \frac{308.15}{298.15}(-56.687) + \left(1 - \frac{308.15}{298.15}\right)(-68.315)$$

$$- \frac{17.995}{10^3}\left(308.15\ln\frac{308.15}{298.15} - 10\right)$$

$$= -56.297 - 0.00298 = -56.300 \text{ kcal/mol}.$$

Equation 21.19 now yields the vapor pressure at 35°C:

$$p_{sat} = \exp\left(\frac{-56.300 + 54.529}{1.987 \times 308.15}10^3\right) = 0.05547 \text{ atm},$$

about 0.05% different from the accepted value. (The ratio of vapor pressures at 35 and 25°C is essentially identical to what would be calculated with Eq. 11.14 and the enthalpy of vaporization obtained from Tables 19.1 and 19.2: $\Delta\tilde{H}_{vap} = -57.796 + 68.315 = 10.519$ kcal/mol).

Example 21.2 Investigate the carbon tetrachloride–methanol system at 50°C with respect to the possibility of the formation of two immiscible phases. Use the Margules coefficients in Table 21.1.

Solution: First plot the molar Gibbs function against the mole fraction of carbon tetrachloride (component 1). (Actually we subtract the values for the pure components since the reference-state values are not relevant to this problem and need not be computed.) The curve in Fig. 21.7 shows a peculiar shape—in particular, a straight line can be drawn so that it is tangent to the curve at two distinct points. The criterion for equilibrium, at a fixed temperature and pressure, of a system containing x_1 moles of CCl_4 and x_2 moles CH_3OH is that the Gibbs function should be minimized. Between the two tangent points on the graph, this can be accomplished if a single phase splits into two phases having compositions corresponding to the two tangent points.

The molar Gibbs function for this system takes the form

$$\tilde{G} = x_1\mu_1^o + x_2\mu_2^o + RT(x_1 \ln x_1 + x_2 \ln x_2) + x_1x_2[A_{12}^* + B_{112}^*(x_1 - x_2)].$$

For the curve in Fig. 21.7 to have two tangent points, there must also be two inflection points, where

$$\frac{d^2\tilde{G}}{dx_1^2} = 0 = \frac{RT}{x_1} + \frac{RT}{x_2} - 2A_{12}^* - 6B_{112}^*(x_1 - x_2).$$

At the incipient point where two phases just begin to appear, these two inflection points join each other, and we also have

$$\frac{d^3\tilde{G}}{dx_1^3} = 0 = \frac{RT}{x_2^2} - \frac{RT}{x_1^2} - 12\,B_{112}^*.$$

Such a point, similar to the critical point in the PVT properties of a pure substance, is called the *consolute point*. Simultaneous solution of the above two equations allows us to develop a criterion for whether two phases can form. At a consolute point, B_{112}^* is related to A_{12}^* by

$$\left(\frac{6B_{112}^*}{RT}\right)^2 = \frac{1}{2(x_1x_2)^3} - \frac{A_{12}^* / RT}{(x_1x_2)^2},$$

where

$$x_1x_2 = \frac{1}{3 + \sqrt{9 - 4A_{12}^* / RT}}.$$

This relationship is shown in Fig. 21.8.

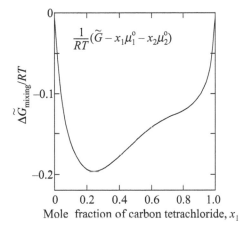

Figure 21.7 Gibbs function change on mixing for binary mixtures of carbon tetrachloride and methanol at 50°C. The shape of the curve can be contrasted with that on Fig. 21.4, where the ideal $\Delta \tilde{G}$ of mixing is also shown.

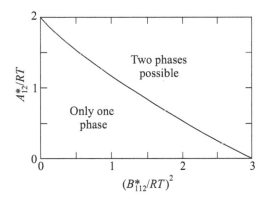

Figure 21.8 Stability of a binary system according to the three-suffix Margules equation.

For additional insight into the special character of this system, plot μ_1 and μ_2 versus x_1 in Fig. 21.9. Actually, we choose to plot $\exp[(\mu_i - \mu_i^\circ) / RT] = x_i \gamma_i$, that is, the *activity* of species i. We should expect μ_1 to increase monotonically with x_1. The fact that it does not is another indication that this system will split into two phases within a certain composition range.

To determine exactly the compositions of the two phases that can coexist, it is expedient to plot μ_1 versus μ_2 (or $x_1 \gamma_1$ versus $x_2 \gamma_2$ as in Fig. 21.10) with x_1 being treated as a parameter. This is because the condition of phase equilibrium, Eq. 16.11, requires the chemical potential of each component to be the same in the two phases. The determination in this case gives $x_1^\alpha = 0.328$ and $x_1^\beta = 0.847$.

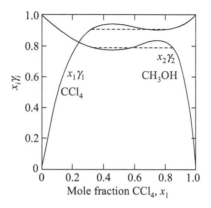

Figure 21.9 Activities of methanol and carbon tetrachloride at 50°C. The two phases in equilibrium are connected by dashed lines.

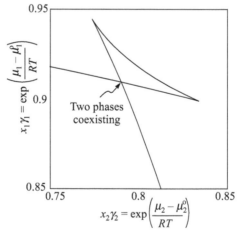

Figure 21.10 Determination of point where the phase-equilibrium condition $(\mu_i^\alpha = \mu_i^\beta)$ is satisfied.

The development in this example should be compared with that in Section 11.2, where the condition of equilibrium of vapor and liquid was explored for a pure substance.

Problems

21.1 a. If the volume change on mixing is identically zero for a binary system, what implications does this have for the Margules coefficients A_{12}^* and B_{112}^* that appear in the expression for the excess Gibbs function

$$\tilde{G}^E = x_1 x_2 [A_{12}^* + (x_1 - x_2)B_{112}^*]?$$

b. If the *excess* entropy of the mixture is identically zero, what implications does this have for the Margules coefficients A_{12}^* and B_{112}^*?

c. If the heat of mixing is identically zero, what implications does this have for the Margules coefficients A_{12}^* and B_{112}^*? Recall that

$$\left(\frac{\partial(G/T)}{\partial T}\right)_{p,n_i} = -\frac{H}{T^2}.$$

21.2 Calculate the enthalpy change on mixing of 1 mol of benzene with 1 mol of cyclohexane at 40°C. The Margules coefficients for this liquid system are given in Table 21.1.

21.3 Put down the Gibbs–Duhem equation in its simplest form (in terms of activity coefficients and mole fractions) for a binary system at constant temperature and pressure. Use this result to show that the following pair of expressions is thermodynamically inconsistent:

$$\ln \gamma_1 = A + (B - A)x_1 - Bx_1^2$$
$$\ln \gamma_2 = A + (B - A)x_2 - Bx_2^2.$$

21.4 Show on the basis of the general three-suffix Margules Eq. 21.2 that the chemical potential of a component can be expressed as

$$\mu_i = \mu_i^o + RT \ln x_i + \sum_j A_{ij}^* x_j + \sum_j \sum_k x_j x_k \left(B_{ijk}^* - \frac{1}{2}A_{jk}^* - \frac{2}{3}\sum_i B_{jki}^* x_i\right).$$

21.5 Show on the basis of Wohl's expansion (Eq. 21.42) that the activity coefficient of a component can be expressed as

$$\ln \gamma_i = \sum_j \frac{A_{ij}}{RT}q_i z_j - \frac{1}{2}\sum_j \sum_k \frac{A_{jk}}{RT}q_i z_j z_k.$$

21.6 Examine some of the van Laar parameters in Table 21.3 to see whether the ratio of parameters seems to be in the ratio of the liquid molar volumes (some of which are given in Table 21.4).

21.7 Show that a ternary system for which van Laar parameters have already been determined for the binary subsystems can be treated by Wohl's expansion method without the necessity for fitting again the binary data if the binary parameters satisfy the relationship

$$\frac{A'_{23}}{A'_{32}} = \frac{A'_{21}}{A'_{12}} \frac{A'_{13}}{A'_{31}}.$$

21.8 For a binary system, derive the van Laar equations from the two-suffix Wohl's expansion in Problem 21.5.

21.9 For a regular solution, the Wohl's parameters are related to the solubility parameters according to $A_{ij} = (\delta_i - \delta_j)^2$.

Show that the two-suffix Wohl's expression for the activity coefficient, given in Problem 21.5, can then be written as

$$\ln \gamma_i = \frac{q_i}{RT}(\delta_i - \bar{\delta})^2$$

where

$$\bar{\delta} = \sum_k z_k \delta_k.$$

21.10 Show on the basis of the general three-suffix Wohl's Eq. 21.42 that the activity coefficient γ_i of a component can be expressed as

$$\ln \gamma_i^* = q_i \sum_j z_j \left(\frac{A_{ij}}{RT} - \frac{1}{2} \sum_k \frac{A_{jk}}{RT} z_k \right)$$

$$+ q_i \sum_j \sum_k z_j z_k \left(\frac{B_{ijk}}{RT} - \frac{2}{3} \sum_i \frac{B_{jki}}{RT} z_i \right).$$

21.11 For a binary system described by the three-suffix Wohl's expansion, the activity coefficients γ_i are given by the expressions (the Scatchard–Harrier equations)

$$\frac{RT}{q_1} \ln \gamma_1 = A_{12} z_2^2 + B_{112} z_2^2 (4z_1 - 1)$$

and

$$\frac{RT}{q_2} \ln \gamma_2 = A_{12}z_1^2 - B_{112}z_1^2(4z_2 - 1).$$

(Recall that $B_{122} = -B_{112}$, and compare Eq. 21.25.) The system of toluene (component 1) and acetic acid (component 2) forms an azeotrope at a pressure of 1 atm, a temperature of 105.4°C, and a mole fraction of acetic acid of 0.627. From these data, determine values of the Wohl's parameters A_{12} and B_{112} at 105.4°C. q_1 and q_2 can be identified with the molar volumes of the pure components (estimated to be $q_1 = \tilde{V}_1^o - 117.6 \text{ cm}^3/\text{mol}$ and $q_2 = \tilde{V}_2^o = 63.3 \text{ cm}^3/\text{mol}$), and the pure-component vapor pressures are estimated to be $p_{1,\text{sat}}^o = 0.859$ atm and $p_{2,\text{sat}}^o = 0.657$ atm at the temperature of 105.4°C.

21.12 a. Does the system discussed in Problem 21.11 exhibit positive or negative deviations from Raoult's law? Would the liquid azeotrope be a *constant-boiling* azeotrope?

b. Evaluate the following rule-of-thumb: "the azeotropic composition always tends to come out in the overhead product in binary distillation; the product at the bottom can tend toward either pure component 1 or pure component 2 depending on where the feed composition is with respect to the azeotropic composition."

21.13 The following data (from Ref. [1]) give measured total pressures for vapor–liquid equilibrium over liquid phases of the stated composition of methylcyclohexane (component 1) and nitroethane (component 2) at 45° C.

x_2	p (bar)	x_2	p (bar)
0	0.15349	0.071	0.19027
0.011	0.16313	0.076	0.19109
0.018	0.17150	0.113	0.19636
0.025	0.17126	0.225	0.20202
0.036	0.17719	0.233	0.20331
0.050	0.18628		

a. Estimate the vapor-phase composition for a liquid-phase mole fraction of $x_2 = 0.15$.

b. Estimate the activity coefficient γ_2 of nitroethane at infinite dilution in methylcyclohexane if the pure-component vapor pressure (of nitroethane) is $p^0_{2,sat} = 0.07855$ bar.

c. Would you expect an azeotrope to form for this system? Why or why not?

21.14 Example 21.3 treats phase equilibrium between two liquid phases of carbon tetrachloride (component 1) and methanol (component 2). The results, at 50°C and based on the Margules expansion of the Gibbs function, can be summarized as follows:

	Phase α	**Phase β**
x_1	0.32816	0.84703
x_2	0.67184	0.15297
γ_1	2.7720	1.0739
γ_2	1.1752	5.1613

Now we wish to add a small amount of acetone (component 3). The Margules coefficients for this system are as follows:

$$A^*_{12} / RT = 2.14 \quad B^*_{112} / RT = 0.38$$

$$A^*_{13} / RT = 0.830 \quad B^*_{113} / RT = -0.115$$

$$A^*_{23} / RT = 0.610 \quad B^*_{223} / RT = 0.091$$

$$B^*_{123} / RT = -0.575$$

a. Assuming that x_3 is very small (and, therefore, does not disturb the values of x_1, x_2, γ_1, and γ_2), calculate values of the activity coefficient of acetone in both phases α and β.

b. Still for the situation where x_3 is very small, work out the value of the distribution coefficient of component 3 between phases α and β, that is, the value of $K = x^\beta_3 / x^\alpha_3$.

21.15 The graphs in Figs. 21.11 and 21.12 were found in the fifth edition of the *Chemical Engineers' Handbook* under azeotropic distillation. Comment on the quality of the sketches, particularly with regard to the partial pressures of the components.

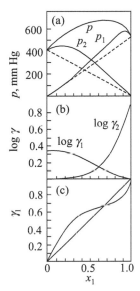

Figure 21.11 Chloroform (1)–methanol (2) system at 50°C. Azeotope formed by positive deviation from Raoult's law (dashed lines). (Data of Sesonske, Ph.D. Dissertation, University of Delaware.)

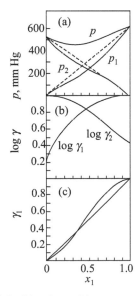

Figure 21.12 Acetone (1)–chloroform (2) system at 50°C. Azeotrope formed by negative deviations from Raoult's law (dashed lines). (Data of Sesonske, Ph.D. Dissertation, University of Delaware.)

21.16 Ethanol (component 1) and water (component 2) form an azeotrope at about $x_1 = 0.895$ at a pressure of 1 atm. It has been suggested that distillation at a lower pressure could avoid the azeotrope. Estimate at what temperature the azeotrope would be shifted to $x_1 = 0.96$. Data on the pure-component vapor pressures are given in Fig. 21.13, and you can assume that the activity coefficients given in Fig. 21.6 for 25°C can be applied at any temperature without correction.

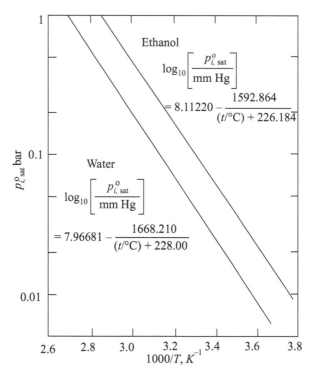

Ethanol

$$\log_{10}\left[\frac{p_{i,\,\text{sat}}^o}{\text{mm Hg}}\right] = 8.11220 - \frac{1592.864}{(t/°C) + 226.184}$$

Water

$$\log_{10}\left[\frac{p_{i,\,\text{sat}}^o}{\text{mm Hg}}\right] = 7.96681 - \frac{1668.210}{(t/°C) + 228.00}$$

Figure 21.13 Vapor pressure for the pure components, ethanol and water.

21.17 A modified van Laar expression for activity coefficients reduces to the following form for a binary liquid mixture:

$$\ln \gamma_1 = \frac{\alpha_{12}^2}{\left(1 + \dfrac{\alpha_{12}^2\, x_1}{\alpha_{21}^2\, x_2}\right)^2} + (x_1 - x_2)\, c_{12} x_2^2 (5x_1 - x_2)$$

and

$$\ln \gamma_2 = \frac{\alpha_{21}^2}{\left(1 + \dfrac{\alpha_{21}^2 \, x_2}{\alpha_{12}^2 \, x_1}\right)^2} + (x_2 - x_1)\, c_{21} x_1^2 (5x_2 - x_1)$$

where

$$\alpha_{ij}^2 = A_{ij} - c_{ij} \text{ and } c_{ij} = c_{ji}.$$

Examine these expressions to determine whether they are thermodynamically consistent. (The values of A_{ij} are not related to Wohl's parameters. They need not be symmetric.)

21.18 If the fundamental equations for the temperature and pressure dependence of the properties of a pure liquid component are

$$\left(\frac{\partial(\mu_i^o / T)}{\partial T}\right)_p = \frac{-\tilde{H}_i^o}{T^2} \text{ and } \left(\frac{\partial \mu_i^o}{\partial p}\right)_T = \tilde{V}_i^o,$$

(where superscript zero denotes the pure state), what equations would apply for the temperature and pressure dependence of the activity coefficient γ_i in a liquid mixture?

21.19 A table of solubility parameters, pure-component vapor pressures, and liquid-phase molar volumes is given in Table 21.4. On the basis of regular-solution theory and the tabulated information, determine whether the system of benzene and cyclohexane should form an azeotrope. If it does, estimate its composition. Does this system show positive or negative deviations from Raoult's law?

21.20 Suppose a compound A_2 dissociates when dissolved in compound B according to the reaction $A_2 \rightleftharpoons 2\,A$. If B and the dissociated species A form ideal solutions (on a mole-fraction basis), what will be the composition dependence of the activity coefficient of A_2, obtained from measurements without regard for the dissociation? The dissociation reaction goes essentially to completion so that there is only a small amount of A_2 in the liquid phase. Would the system exhibit positive or negative deviations from Raoult's law? What problems arise if we try to use the unsymmetric convention for activity coefficients?

21.21 Discuss the relationship between a thermodynamic consistency test based on Eq. 16.4 and one based on the use of the Gibbs–Duhem equation. You may confine your attention to a binary system. Do not hesitate to be quantitative, using equations and derivations.

21.22 At 25°C, a binary system containing components 1 and 2 is in a state of three-phase vapor–liquid–liquid equilibrium. Analysis of the two equilibrium liquid phases (α and β) yields the following compositions: $x_2^\alpha = 0.05$ and $x_1^\beta = 0.05$. Vapor pressures of the two pure components at 25°C are:

$$P_{1,\text{sat}}^o = 0.65\,\text{atm} \quad \text{and} \quad P_{2,\text{sat}}^o = 0.75\,\text{atm}.$$

You can use low-pressure approximations for this system. The only other approximation you can use is that the system is fit by the three-suffix Margules equation—one involving $A_{1,2}^*$ and $B_{1,1,2}^*$. Do not assume that $\gamma_i = 1$ for the major components. Determine

a. The Margules coefficients $A_{1,2}^*$ and $B_{1,1,2}^*$.

b. The activity coefficients $\gamma_1^\alpha, \gamma_2^\alpha, \gamma_1^\beta$, and γ_2^β for components 1 and 2 in the equilibrium α and β liquid phases.

c. The equilibrium pressure p.

d. The equilibrium vapor composition y_1.

21.23 In the low-pressure approximation, the vapor pressure of a pure component obeys equations strongly similar to those describing the temperature dependence of the equilibrium constant $K(T)$ for a chemical reaction.

a. What equation relates the pure-component vapor pressure to the secondary reference state quantities used for the liquid and ideal-gas states?

b. What equation, therefore, follows for the derivative with respect to temperature of the pure-component vapor pressure?

c. If the temperature dependence of the ideal-gas heat capacity can be expressed as

$$\tilde{C}_{pi}^* = a_i + b_i T + c_i / T^2 + \gamma_i T^2$$

and a similar expression applies to the liquid, what equation should result for

$$\ln\left(\frac{p_{i,sat}^{o}(T)}{p_{i,sat}^{o}(T_o)}\right)?$$

Show some basis for your result.

21.24 It is an old saying that "A watched pot never boils." But we know that this is not true. An old pot watcher has come up with a more penetrating observation. He states that a pot boiling at constant pressure always gets hotter—or at least that it never gets cooler. Discuss the basis for such a statement in terms of your knowledge of the thermodynamics of binary vapor–liquid equilibrium.

21.25 Cite advantages and disadvantages of fitting liquid-phase activity coefficients and the excess Gibbs function by means of the following three methods:

a. Regular-solution theory.

b. The three-suffix Margules equation.

c. The Wilson equation.

21.26 Data for the enthalpy of mixing of water (component 1) and ethanol (component 2) are shown in Fig. 21.14. Discuss how to use the data in the figure to obtain activity coefficients at 35°C from those shown in Fig. 21.6 at 25°C.

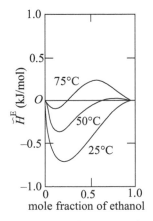

Figure 21.14 Excess molar enthalpy of mixtures of water (1) and ethanol (2).

21.27 Data for the enthalpy of mixing of water (component 1) and ethanol (component 2) are shown in Fig. 21.14. Discuss how such data might have been calculated on the basis of the van Laar coefficients given in Table 21.3.

21.28 Data on the activity coefficients of six binary systems are reported as shown in Fig. 21.15, plotted against mole fraction. Which of the results are reasonable? Explain your reasoning.

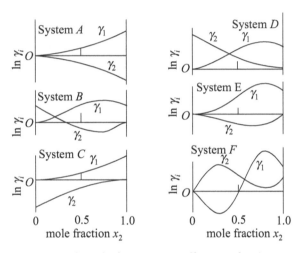

Figure 21.15 Reported results for activity coefficients of six binary systems.

21.29 The Wilson equation is to be compared with the regular-solution theory by answering the multiple-choice questions below with one of the following answers:

1. Neither   2. Wilson equation
3. Regular-solution theory   4. Both

   a. Parameters can be predicted from pure-component data alone.

   b. Parameters can be based on binary vapor–liquid equilibria at a single temperature.

   c. Theory readily gives predictions over a range of temperatures.

   d. Data on ternary mixtures can be used to refine the data fit.

   e. The theory applies readily to multicomponent systems.

   f. The equations can describe an azeotrope in a binary system.

g. The theory can describe limited liquid miscibility in a binary system.

21.30 Vapor–liquid equilibria in the system of 2-propanol (component 1) and toluene (component 2) at 80.6°C show an azeotrope at a composition of $x_1 = 0.77$ and a pressure of 1 atm. Pure-component vapor pressures at this temperature are estimated to be 0.923 atm for 2-propanol and 0.390 atm for toluene. Estimate Henry's law coefficients (on a mole-fraction basis) for both components in each other. For example, for dilute solutions of component 2 in component 1, Henry's law takes the form $p_2 = H_{2(1)}x_2$ and $H_{2(1)}$ is the sought quantity.

What value would $H_{2(1)}$ have if the systems were ideal?

References

1. Robert van Orye. *Thermodynamics of Binary and Multicomponent Mixtures of Hydrocarbons and Polar Organic Liquids.* Dissertation. University of California, Berkeley, 1965.

2. Michael J. Holmes and Matthew van Winkle. "Prediction of ternary vapor-liquid equilibria from binary data." *Industrial and Engineering Chemistry*, **62**, 21–31 (1970).

3. I. Mertl. "Liquid-vapour equilibrium. L. Prediction of vapour-liquid equilibria from the binary parameters in systems with limited miscibility." *Collection of Czechoslovak Chemical Communications*, **37**, 375–411 (1972).

4. J. M. Prausnitz. *Molecular Thermodynamics of Fluid-Phase Equilibria* (Englewood Cliffs, New Jersey: Prentice-Hall, 1969).

5. Grant M. Wilson. "Vapor-liquid equilibrium. XI. A new expression for the excess free energy of mixing." *Journal of the American Chemical Society*, **86**, 127–130 (1964).

6. Wendell M. Latimer, *The Oxidation States of the Elements and Their Potentials in Aqueous Solutions* (Englewood Cliffs, New Jersey: Prentice-Hall, 1952).

Chapter 22

Surface Systems

We touched on surface systems in Chapter 10, mainly to provide an example of something besides PVT systems in thermodynamics. It should be clear that other systems can be taken through a Carnot cycle and provide a basis for the thermodynamic temperature scale. The contrast showed that the strong coupling between temperature and mechanical effects make PVT systems preferable for heat engines and refrigerators.

However, surface systems have their own interest in multicomponent thermodynamics. The surface tension can be changed by adsorption of chemicals at the interface between phases and lead to surface-tension-driven flows. Noteworthy here is the Gibbs adsorption equation

$$d\sigma = -s^\sigma dT + \tau dp - \sum_i \Gamma_i d\mu_i, \qquad (22.1)$$

where s^σ, τ, and Γ_i are surface excesses of entropy, volume, and moles of species i. σ is the surface-excess Gibbs energy, and is called the surface tension of the interface. The surface excesses mean that the extensive properties of a system are allocated first to the two adjacent bulk phases, and what ever is left over is assigned to the interface. There is ambiguity as to where the interface is located. Gibbs took it to be a surface somewhere in the interface, whose exact position is initially arbitrary. The ambiguity is taken away

The Newman Lectures on Thermodynamics
John Newman and Vincent Battaglia
Copyright © 2019 Jenny Stanford Publishing Pte. Ltd.
ISBN 978-981-4774-26-0 (Hardcover), 978-1-315-10861-2 (eBook)
www.jennystanford.com

by, for example, assigning zero to a particular excess quantity. This could be the adsorption of a solvent or the net mass of the interface. Something which does not change when the position of the defining plane is changed can be called a Gibbs invariant. For example, the surface tension of the interface is a Gibbs invariant.

With Gibbs's method, no volume is assigned to the interface, and τ is zero. The corresponding term drops out of Eq. 22.1. This is a good reminder that the number of independent thermodynamic variables decreases by one when the number of phases increases from one to two. Thus, for a two-phase system, the temperature specifies the pressure, and the pressure cannot be set independently.

The above equation is Eq. 7.14 in Ref. [1], where it is discussed more extensively. There is also a derived equation, called the Lippmann equation, for electrified interfaces (Eq. 7.26 in Ref. [1]).

A full treatment of multicomponent surface systems is not given here. Instead we describe some interesting phenomena related to the subject and try to make clear why the subject is important.

Certain chemical-separation processes can be devised which rely on adsorption at an interface. By bubbling air through a solution, surface-active agents can be adsorbed at the interface and thereby be removed from a solvent. Activated charcoal is used in gas masks to take out toxic gases.

Competitive adsorption of chemicals at an interface can have a profound effect on heterogeneous reactions. The surface may then act as a catalyst, greatly enhancing the rate of a reaction without changing the equilibrium composition. Catalysis is part of the subject of chemical kinetics, but it is vital to understand the underlying thermodynamics before delving into the details of catalysis, or heterogeneous reaction kinetics. Some electrocatalytic phenomena are discussed in Chapter 8 of Ref. [1].

Equation 22.1 introduces the surface tension. This provides an important means for measuring the surface concentration Γ_j. The amount of a species adsorbed at the interface is generally quite small, and one needs a material with a very high surface area to obtain the value by analytical chemistry. Equation 22.1, on the other hand, permits one to obtain an accurate value by measuring the surface tension. To get the surface-excess entropy, one can measure the temperature dependence of the surface tension. As you have learned, measuring the entropy is generally difficult because it requires a reversible mixing experiment. The Gibbs adsorption equation provides such a reversible mixing experiment in a creative way.

When the surface tension varies, a tangential force is exerted on a fluid interface (in addition to the pressure effect of the perpendicular force discussed in Chapter 10). This leads to surface-tension-driven flow in the motion of drops and to electrocapillary motion, discussed in Ref. 1.

These many interesting aspects of surfaces we leave to other courses on separations, chemical reactor design, and electrochemistry.

Reference

1. John Newman and Karen E. Thomas-Alyea, *Electrochemical Systems* (Hoboken, New Jersey: John Wiley and Sons, Inc., 2004).

Appendix

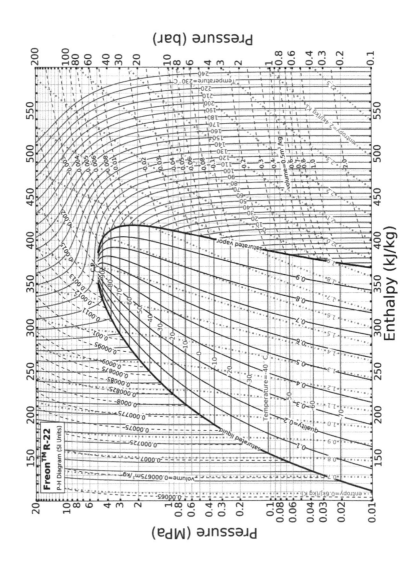

Figure A.1 *p–H* diagram of chlorodifluoromethane (Freon 22 refrigerant). The graph was created using the equation of state published in Ref. [1] and using the IIR reference state of enthalpy = 200 kJ/kg and entropy = 1 kJ/kg-K for saturated liquid at 0 °C (Graph courtesy: Chemours Fluorochemicals Technology Group).

Reference

1. A. Kamei, S. W. Beyerlein, and R. T. Jacobsen, "Application of nonlinear regression in the development of a wide range formulation for HCFC-22," *International Journal of Thermophysics*, **16**, 1155–1164, 1995. doi: 10.1007/BF02081283

Index: General

Index: Chemical Compounds